APR  2013

W9-CCD-908

# Date Due

| | | | |
|---|---|---|---|
| | | | |
| | | | |
| | | | |
| | | | |
| | | | |
| | | | |
| | | | |
| | | | |
| | | | |
| | | | |
| | | | |
| | | | |
| | | | |
| | | | |
| | | | |
| | | | |
| | | | |
| | | | |

# Science at the Frontiers

# Science at the Frontiers

## Perspectives on the History and Philosophy of Science

### Edited by William H. Krieger

LEXINGTON BOOKS
*Lanham • Boulder • New York • Toronto • Plymouth, UK*

Published by Lexington Books
A wholly owned subsidary of The Rowman & Littlefield Publishing Group, Inc.
4501 Forbes Boulevard, Suite 200, Lanham, Maryland 20706
http://www.lexingtonbooks.com

Estover Road, Plymouth PL6 7PY, United Kingdom

British Library Cataloguing in Publication Information Available

**Library of Congress Cataloging-in-Publication Data**

Science at the frontiers : perspectives on the history and philosophy of science / edited by
William Krieger.
    p. cm.
  Summary: "'Science at the Frontiers: Perspectives on the History and Philosophy of
Science' brings new voices to the study of the history and philosophy of science. It
supplements current literature on the history and philosophy of science, which is often
focused on the philosophy of physics, by highlighting sciences that are overlooked by the
current literature and by viewing classic problems in the field from new perspectives.
William H. Krieger, himself an archaeologist and philosopher of science, brings together
scientists, philosophers of science, and historians of science to write on the lessons that
the field stands to learn from case studies in such disciplines as archaeology, medicine,
biology, and others. These essays answer many of the questions that have resisted
solution in the classical canon while raising new questions born out of new perspectives
on the history and philosophy of science. Those studying the philosophy and history of
science and those who are already practicing scientists, philosophers of science, and
historians of science will gain a great deal from these essays"—Provided by publisher.
  Includes bibliographical references and index.
  ISBN 978-0-7391-5014-6 (hardback)—ISBN 978-0-7391-5015-3 (paper)
  1. Science—Philosophy. 2. Science—History. I. Krieger, William, 1969-
  Q175.3.S3175 2011
  501—dc23
                                                          2011024218

# Table of Contents

# INTRODUCTION:
# Defining Frontiers in the History and Philosophy of Science

## William H. Krieger

In summer 2003 a group of scientists, historians, and philosophers of science met in Pittsburgh for an NEH summer institute on Science and Values. When I signed up for this program, I thought this would be a good venue to explore these issues, and I was certainly not disappointed on that score. However, as the summer progressed, I found myself focusing more and more on something incidental to the organizers' program, the variety of scientific disciplines and approaches that were being brought to bear on a number of hot issues in the history and philosophy of science (HPS), and on the relationship between these fields and scientific practice.

This excitement led me to think back to the history and philosophy of science courses and texts that I used in my own education, both as a graduate student, and as the "continuing education" that is the life of an academic. In the course of this self-study, I realized that the anthologies available to students, scholars, and interested science junkies alike largely ignore what I believe is interesting and exciting in this dynamic field. Our focus on a small handful of case studies and on a small number of core HPS disciplines means that we come away from our literature believing that there are a limited number of problems for people in HPS to focus on, a limited number of fields from which to draw data, and therefore, a limited number of lines of reasoning from which to find solutions. Whether due to the quirks of history that formed HPS as a field or due to more modern agendas, HPS collections are not forcing ourselves to ask questions about the sciences and topics that make up that canon.

As a philosopher of science who has also been a working archaeologist since the early 1990s, I am aware of this issue on more than an academic level. Many of the problems that are at the heart of the philosophy of science look very different from an archaeological perspective. By bringing in case studies and perspectives from my 'outsider' science's perspective, I am able to bring new tools and ideas to both the philosophical and archaeological tables.

Although I believe that expanding the palate is a good thing, I have, on more than one occasion (and at more than one professional meeting) run into traditionalists who have questioned not my conclusions, but the idea that my field should have a role to play in discussions of HPS. Many of my colleagues, people who have similar interests in the state of our field, have had similar experiences, and these thoughts and conversations led me to begin to send out a call for authors, asking them the question: "could we do this another way?" Can we think about HPS from the perspective of sciences that do not get much time in the HPS spotlight, and can we use our talents to focus on classic problems in HPS in a different way or from a different perspective?

This book is designed to introduce advanced undergraduate and graduate students, as well as people interested in the theoretical underpinnings of science, to HPS from a variety of new perspectives. The authors are scientists, historians of science, or philosophers of science, representing a wide variety of positions and disciplines. Each author was asked to write a chapter representing an issue of interest to them from their frame of reference and field of expertise. These chapters, each written specifically for this book and this audience, make this an interesting and unique text for people interested in HPS. Rather than dividing the book up by discipline or by specific HPS problem (as is the tradition), I selected pieces for this book that spoke to the two themes referenced above: learning something new from the history of our field (section 1), and looking to new fields and problems (section 2).

# SECTION 1: ACTUALLY LEARNING FROM HISTORY

In "Rhetoric and Faith in the Hippocratic Philosophy of Medicine," *Adam Roth* questions the status quo in a case study in the founding of Western medicine, which reportedly happened in the fifth century BCE. According to most scholars, the birth of this discipline occurred due to a stark division between the magico-religious healing practices characteristic of the Archaic Period, and the empirical-rational pursuit of medicine of the Hippocratics. Drawing on materials from the period and his knowledge of the history of medicine, Roth proposes that Hippocratic physicians remained well within and committed to the religious and cultural confines of the community, including its belief in divine healing. Since Hippocrates and his followers arrive on a medical scene that is already populated by a wide variety of venerable healers-herbalists or root cutters who sell and administer medicinal plants, gymnastic trainers, midwives, religious practitioners of temple medicine—they compete with other healers for credibility and for clients, and follow—in part, for their own livelihood—culturally entrenched beliefs in the divine. Evidence from the Hippocratic Corpus suggests that the Hippocratics rely on rhetoric to craft an identity for themselves vis-à-vis other medical healers of their time.

In "The Wisdom in Wood Rot: Finding God in Early-Modern Scientific Explanation," *Eric Palmer* notes that we tend to explain designed objects by reference to their purposes, and philosophers of science and scientists themselves tend to accept this kind of explanation as legitimate. This is most obvious in our time in fields such as archaeology: a comprehensive explanation for why Mayan temples are just as they are will include reference to the intentions of their designers. But reference to purposes in natural science reaches beyond the human sciences, and back to the beginnings of science. It received particularly careful attention in the early eighteenth century in what we might consider to be some of the earliest recognizably modern philosophy of science. In the general topic of natural theology, tracing God's purposes in the Creation was given a particularly large boost in prominence in England in the late seventeenth century by Isaac Newton and Robert Boyle. Their encouragement led, respectively, to the famous Leibniz-Clarke debate, which concerns the intersection of theology with physics and cosmology, and to a vogue in physico-theology, in which God's purposes were traced in biology and geology. Palmer focuses especially on the broad and amusing variety of proposed explanations, two samples of which are reflected in its title. He also considers philosophical debates and tests of the legitimacy of these two sorts of explanation. Both philosophers and scientists, including Gottfried Leibniz, Samuel Clarke, François Fénelon, Benedict Spinoza, Elie Bertrand, Noel Pluche, and Voltaire, are at the center of the discussion.

Also interested in changing our ideas about the modern period, *Anya Plutynski*, in "In Defense of "Rationalistic" Science," focuses our attention on one of the best-known (if not best understood) figures from this period. In the last four-five sections of his Principles of Philosophy, Descartes essentially offers up a set of probabilistic, inference to the best explanation-style arguments for his physics, and (effectively) introduces the problem of empirical equivalence. Positing that Descartes was a much more subtle thinker than many give him credit for, Plutynski engages in a bit of "myth-busting" about Descartes' epistemology, and philosophy of science, providing a nice prompt and new perspective for discussing the difference between scientific realism, empiricism, and rationalism.

In "Lost Science, Deepwater Shipwrecks, and the Wheelbarrow of Archimedes," *Bridget Buxton* showcases the role that the emerging science of deep-sea underwater archaeology can and should play in the recovery of knowledge of ancient scientific achievements. During the Hellenistic period, a variety of cultural forces caused Hellenistic scientists to focus their energies on applied science and technology, causing these fields to reach a level of sophistication not seen again in Europe until the industrial revolution. However, although we are told that other incredible scientific achievements were occurring during this period (from a number of textual sources), the focus on applied science and technology in the preserved historical record, and a lack of archaeological evidence of other scientific achievements, means that we know comparatively little about the practical scientific achievements of the ancient Greeks and Romans, and that we have been less inclined to believe accounts of their more advanced and unique innovations. According to Buxton, underwater exploration is now poised to change that picture. The Antikythera mechanism, a remarkably sophisticated 2000-year-old astronomical analog "computer" discovered on a Roman shipwreck, is considered here as an example of lost high technology that survived only in this unique archaeological environment. Discoveries such as this not only draw attention to our inadequate understanding of the evolution of scientific thought in western history, but also beg the question of how an intellectually liberal and prosperous civilization could turn down the path into scientific regression, intellectual persecution, and Dark Age. Deepwater shipwrecks, as well-preserved time capsules that both embodied and transported the scientific achievement of the Classical world, have an unequalled archaeological potential for improving our understanding of what our cultural forebears found and lost—and why.

In "Removing the 'Grand' from Grand Unified Theories: An Archaeological Case for Epistemological and Methodological Disunity," *William H. Krieger* uses a case study from archaeology to unearth the rationale behind the archaeological push toward scientific unification. In the 1960s, processual, or new archaeology, based on Hempelian explanation, proposed a roadmap for Archaeology to become a science. Regardless of contemporary archaeologists' feelings about the successes or failures of new archaeology, many archaeologists, like many scientists in general, continue to believe in one of its tenets, scientific (here meaning both theoretical and methodological) unification. A grand unified theory of archaeology would trickle down, standardizing methodology and practice. Krieger argues that, for archaeology at least, the position that a unified theory should determine local scientific goals and practices would be detrimental. Further, in presenting this study, the author focuses on the role local issues have historically played (on an unconscious level) on archaeology's theoretical development, arguing that these features must (consciously) continue to guide the field, resulting in a multitude of local archaeologies, and not in a new grand unified archaeological theory. In short, the science of archaeology presents the philosophical community an example of a field that will only be able to do good science if it can resist this sort of scientific unification. By understanding the history (both archaeological and philosophical) of the new archaeology and its critiques, Krieger argues for a more complicated and more accurate scientific foundation to archaeology.

In "The Virus As Metaphor In American Popular Culture, 1967-2010," *Steven Hatch* asks how well our understanding of the world has kept up with the incredible medical breakthroughs of the last century. Since the advent of variolation—the deliberate introduction of smallpox virus into a human host in order to prevent full-blown disease later—we have slowly but progressively moved away from interpreting human epidemics as supernatural phenomena. With the scientific triumphs of cell theory, the establishment of Koch's postulates, the use of antimicrobial chemotherapy and the creation of modern vaccines, fewer and fewer people are likely to see the Hand of God in infectious epidemics, a view that was commonplace among even educated people until within the past 200 years. Clearly, we live in a more enlightened age, having conquered such antiquated notions, living in a brave new world where we maintain an enlightened view of epidemics, where a virus is just a virus trying to make its way in the world and nothing more. Or is it? Do we really understand infectious epidemics as they are or do we have more in common with our ancestors who turned viruses into metaphors—merely exchanging new acceptable metaphors for old discarded ones. Hatch focuses on the popular literature on several viral epidemics over the past three decades to demonstrate that indeed we are guilty of "metaphorizing" virus. However, unlike the previous over-readings of viral epidemics (which had a decidedly fatalistic bent, at least in the West) our modern readings are highly politicized and frequently take two forms: one of nationalism and a second of "nature's revenge." Such metaphors often feed into, and are fed by, right-wing and left-wing ideologies, respectively. Just as in days of yore, these metaphors serve to muddle our understanding of the biology of viruses, and as a consequence cheapen both our politics and our understanding of science.

In "Gender, Germs, and Dirt," *Sharyn Clough*, also interested in social forces adversely affecting (or infecting) human health, shows that a better understanding of history and social factors answers a seemingly bizarre finding that otherwise leads to a gender-based problem with the widely accepted "hygiene hypothesis." This hypothesis addresses the well-established correlation between increased cleanliness and sanitation, and increased rates of asthma and allergies. Recent studies have added to the scope of the hypothesis, showing a link between decreased exposure to certain bacteria and parasitic worms, and increased rates of depression and intestinal inflammatory diseases, respectively. What remains less-often discussed in the research on these links is that, paradoxically, women have higher rates than men, of asthma, allergies, auto-immune and inflammatory disorders generally, and also depression. Clough argues that feminist political values can usefully augment the hygiene hypothesis, specifically by highlighting the role that the differential gender socialization of boys and girls might contribute to the hygiene picture. That standards of cleanliness are higher for girls than boys, especially under the age of five when children are more likely to be under close adult supervision, is a robust phenomenon. Adding a gender analysis to the hygiene hypothesis greatly strengthens the explanatory force of the latter. Documenting the link between exposure to healthful bacteria and parasites on the one hand, and children's gender role development on the other, requires an interdisciplinary approach attentive to immunological research as well as the research of feminist sociologists, and anthropologists. Clough reviews these various literatures and highlights the fruitful avenues for further research, both immunological and epidemiological, that are made visible once gender socialization is taken into consideration.

# SECTION 2: THE OTHER SIDE(S) OF SCIENCE

Recently, a number of prominent scientists and scientifically oriented philosophers have come out in favor of atheism with respect to the existence of God and the supernatural. Other scientists have come out in support of Christian views of God. In "The Agnostic Scientist: The Supernatural and the Open-Ended Nature of Science," *Brian L. Keeley* questions whether these two polar positions are (as people on both sides of the a/theistic chasm have asserted) the only appropriate scientific (and rational) positions that can be held. Going further, he questions whether a categorical claim on the existence or non-existence of a hypothesized entity (or class of entities) is consistent with an open-ended and always-revisable characterization of science. After all, as is often proclaimed, hypotheses are never "proven" by science and every scientific conjecture is, at least in principle, open to future revision on the basis of new observations and theoretical innovation. This understanding of science would appear to advocate for religious *agnosticism*, not atheism. This chapter will explore this apparent tension and attempt to understand whether scientists *qua* scientists must take a particular stand on the existence of the supernatural.

*Yuri T. Yamamoto* focuses on scientific responsibility by analyzing the tension between a research scientist's two opposing goals: the need to help make policy and the need for personal academic success. In "Participating in a Contentious Natural Resource Debate as a Scientist: For Better or For Worse," Yamamoto argues that conflicting goals and standards between these two foci of science create an obstacle for successful integration of scientific practice and policy, an obstacle that she believes can be overcome by properly engaging stakeholders outside of the academic community. This article examines what happens to science under intense citizen involvement in a contentious natural resource policy debate based on archival documents and interviews of academic scientists who participated in the North Carolina Wood Chip Production Study. The goal of the extended peer review community of citizens and scientists in this study was to enhance policy-related research by ensuring that researchers would ask relevant questions and incorporate local knowledge and values into the study. However, to accomplish this goal, Yamamoto shows that academic scientists were sometimes required to stray away from the standards set by their academic community in ways that resulted in the creation of good science policy. Yamamoto argues further that citizen involvement may also expose untested assumptions and gaps in academic research, leading to more robust results, new areas of research, and additional funding.

Also interested in the ways that science and science policy is made, *Lawrence Souder* is interested in situations when scientists step into the role of peer reviewers in "Conflicting Values in Post-publication Disputes: The Case of Transgenic DNA in Mexican Maize." Scientific research communities traditionally have tried to dissociate social values from constitutive values. In its quest for objective knowledge, however, the practice of science cannot avoid social values because the dialogical process of reaching rational consensus is inherently social. The peer review process in particular is informed by social as well as constitutive values. This process is complicated by the fact that this process does not always cease at publication. Occasionally post-publication errata protract the peer review process. In such cases the wider scientific community enters the dialogue and the values underlying the disputes become more variegated. The analysis of a case study suggests that an unnecessary emphasis on the value of impartiality can cause a dispute over the validity of research to devolve into useless conflicts over integrity. Souder argues that this conclusion reinforces the importance of the separation of social from constitutive values in the practice of science.

Rather than argue that social values should be more thoroughly separated from constitutive values in "On Scientific Advocacy: Putting Values and Interests in Their Place," *Evelyn Brister* argues for an expanded role for scientists as policy makers. Twentieth century philosophy of science supported a model of science as a logical enterprise, free from value judgments. Recent critics have shown that nonepistemic values have played a role in the development and testing of even our best scientific theories. In light of this criticism, how can a recognition and reevaluation of values in science affect scientific practice? Would it permit scientists to act as advocates for policies within their role as scientists? A role for scientists as advocates is typically limited to advocacy for improvements in science education and advocacy through professional organization for the sake of advancing science. Given these questions, Brister asks whether the role of advocate be expanded in light of recent analyses of the positive role that values play in science? Brister uses a case study in conservation biology, a field that actively questions how values shape research agendas, which is especially important, as this field that is dedicated to a normative goal: preserving biodiversity. Nonetheless, there are local struggles over the limits of normativity, such as whether advocacy is appropriate in some cases but not in others. Brister argues that scientists can legitimately act as advocates by identifying research questions that are salient from a policy perspective and by working toward interdisciplinary collaboration for the sake of problem-solving.

Another way that scientists' actions can affect science is in the review process. In "Bias, Impartiality, and Conflicts of Interest in Biomedical Sciences," *Kristen Intemann* and *Inmaculada de Melo-Martín* argue that most scientists erroneously buy into the assumption that social values may legitimately play a role in decisions about what to fund or how to apply scientific knowledge, but not in the context of conducting basic research or clinical trials. Using examples from a variety of biomedical disciplines (including epidemiological research on health disparities, the development and testing of the HPV vaccines, clinical trials on anti-depressants, and research on cesarean births) Brister and Melo-Martín argue that the social aims of research are relevant to decisions throughout the research process. As a result, researchers must identify and endorse particular social aims at stake in research, as well as how those aims might best be promoted by the methodological options available. Further, the authors show that identifying and distinguishing the ways that social values can be relevant to scientific decision-making helps provide insight into what practices might help facilitate the identification, evaluation, and adoption of social values at stake in biomedical research. In particular, they argue that while greater training in ethics for researchers is desirable, it will not be sufficient because biomedical researchers cannot be expected to have all of the relevant expertise needed. More substantive collaboration with social scientists and ethicists throughout the research will also be necessary.

While most believe that, problem riddled or not, at least science has the ability to police (and legislate) itself, *Glenn Sanford* shows that this in not always the case. In "Science, Religion, and Duty in Parenting Choices," Sanford shows just how little control scientists have over matters that would seem to be clearly within their spheres of expertise and influence. In his case study on childhood vaccinations, Sanford analyzes the process that has led to the United States' position that allows people to refuse to vaccinate their children. He shows that this process, keen to protect the interests of a number of stakeholders, has almost totally ignored the scientific community, and in doing so, may have ignored the health of the populace in general, arguably a stakeholder that should have been at the table.

# CONCLUSION

As noted above, I made the decision to eschew traditional groupings (by philosophical problem or by discipline), instead focusing on ways that materials in HPS can get us to think differently about our field. This is in keeping with my idea for the book. When I commissioned 'Frontiers,' my goal was to generate new case studies for people interested in HPS to talk about. I wanted to expose people in the field to other voices, to people working on their problems from different perspectives, and to look at the ways that the philosophy of science connects to the world of scientific practice.

Although *Science at the Frontiers* asks the field to consider the benefits of asking new questions, I hope that this book can help answer questions that have resisted solution in the classical canon as well, covering the field in a way that will challenge the reader to think from a new perspective. To this end, this book has chapters on traditional issues such as: explanation and confirmation (2, 5), science and values (5, 10, 11, 12, 13), feminist science (7, 11, 12), science and religion (1, 2, 8, 13), and science education (1, 2, 3, 6, 7, 9, 10, 11).

Finally, one of the largest complaints that I used to get from my students and from science enthusiasts is that the readings in traditional HPS texts were almost all written by dead people. As each of the readings in *Frontiers* are new, this book will give the reader a real feel for regions where philosophical and historical boundaries (and buttons) are being pushed. Whether it is used for a course in HPS, as a research tool for someone working in the field, or as a way for the science enthusiast to see what is out on the scientific frontiers, this book should be seen as both a companion and a challenge, to the canon. I (along with my authors) welcome you to our view of an exciting (and we would argue, dynamic) field, the history and philosophy of science.

# SECTION 1: ACTUALLY LEARNING
# FROM HISTORY

# Chapter 1

# Rhetoric and Faith in the Hippocratic Philosophy of Medicine

## Adam D. Roth

Historical accounts of Hippocrates repeatedly render him legendary for being the first physician to abandon magico-religious medicine—including divine explanations for disease and magical means for curing them—in favor of a conception of medicine that was secular, natural, and, arguably, proto-scientific—prescient of the hard sciences to come. In part on the basis of this depiction, we are led to believe that—once revealed—Hippocrates' theories and practices were readily accepted by members of the Greek community; that the magico-religious medicine which dominated the Archaic Period quickly gave way to the self-evident truths discovered through the natural, empirical, and rational pursuit of medicine advanced by the Hippocratics[1] in fifth and fourth century BCE Greece.

As this chapter will show, however, the rise of Hippocratic medicine in ancient Greece involves a more complex and nuanced story than traditional histories generally reveal. Unlike scholars who posit a radical distinction between Archaic and Classical conceptions of medicine, I will show how the Hippocratics—the epitome of Classical medicine and the progenitors of the Western medical tradition—were implicated by, and in many ways committed to, the traditional and religious medical beliefs adopted by their communities—in all their divine and supernatural manifestations. Furthermore, the Hippocratics emerged onto a medical marketplace that was already populated by a wide variety of respected healers—herbalists or root cutters who sold and administered medicinal plants, gymnastic trainers, midwives, religious practitioners of temple medicine, to name a few—they competed with other healers for credibility and for clients, and abided by—in part, for their own livelihood and credibility—culturally entrenched and axiomatic beliefs in the divine.

If they were not able to differentiate themselves purely on the basis of their (perceived lack of) commitment to religion, the *Hippocratic Corpus*[2] suggests that Hippocratic physicians relied on their oratorical and literary skills, on rhetoric, to craft an identity for themselves vis-à-vis other medical healers of their time. Specifically, I will argue in this chapter that the Hippocratics sought to position themselves as a viable alternative to other healing practitioners by engaging in rhetoric to promote and define their art, to secure and advance their social standing in society, and to defend and fight against attacks that render suspect their practices of medicine.

My argument in this chapter will unfold as follows. After a brief overview of the social context of healing in fifth and fourth century BCE Greece, I will identify the major obstacles that the Hippocratics encountered: first, the socially sanctioned medical practices of sacred healers against whom the Hippocratics had to compete; second, their low status in society as craftsmen rather than as experts; finally, the spread of charlatans and imposters whose claims to medical knowledge made it difficult to discern the Hippocratics from them. I will next explore how the Hippocratics reached to the art of rhetoric as a way to overcome these obstacles, namely, their efforts to differentiate themselves from other healing practitioners, to establish their own standing in society as intellectuals, and to distance themselves from impostors.

## Sociology of Healing in Ancient Greece

One of the most significant challenges facing modern scholars of ancient medicine is to account for the concurrent rise of Hippocratic medicine and its naturalistic theories of disease and practical means for curing them, and the growth of the healing cult of Asclepius and other healing divinities. I concur with G. E. R. Lloyd when he suggests that

> Earlier positivist historians of medicine like to represent the former as superseding the latter. First there was religious medicine, the type represented in the shrines of Asclepius at Epidaurus, Athens, Cos, and elsewhere. Then came naturalistic accounts of disease and their cures. Science, in a word, on this view, overtook religion as the basis of medical practice.[3]

Evidence reveals a more nuanced history; the growth and rise of temple medicine occurs roughly around the same time that the Hippocratics appear in the fifth century; temple healing exists side-by-side with Hippocratic medicine. Furthermore, medico-religious shrines at Epidaurus and Pergamum testify to the prevalence of magico-religious medicine well into the Hellenistic era. In short, traditional and religious conceptions of medicine were not extinguished by Hippocratic medicine, and Hippocratic texts show that even they resorted to divine prayers and outwardly pious gestures.[4]

The co-presence of traditional and Hippocratic medicine produced a competitive marketplace for healers, especially for the Hippocratics, since they were newcomers to the scene and because their ideas and methods were unconventional. Although the Hippocratics are often regarded by historians of medicine as a popular and "rationalizing" force in the medical scene of the time, the fact of the matter is that their naturalistic ideas about medicine, their attribution of disease to organic causes, and their practical, hands-on treatments, were competing with a venerable enterprise of magico-religious medicine that believed in the divine origin of disease and magical means for healing them. Since Greek society in the fifth century placed much value on tradition, the Hippocratics found themselves from the start in a disadvantageous position. This is why Julie Laskaris proposes that traditional (mainly religious) medicine stood in the way of the Hippocratics, given:

> the venerability of traditional medical concepts and practices; their occasional efficacy, real and perceived; and their close ties with religion and magic, [which] made them forces with which secular practitioners had to contend intellectually and economically.[5]

Religious healers were respected members of the community because they were socially sanctioned and long immersed into the fabric of Greek life. Hippocratic medicine, on the other hand, had to fight for its respective place in the medical community and for the confidence and trust of Greek citizens who were its potential clients. Hui-hua Chang argues that "because the Greek society in the Classical period was conservative, religious, and superstitious in general, it is hard to believe that the populace would greatly appreciate the novel medical theories of rational doctors," and that the community was skeptical of the Hippocratics' unconventional practices.[6]

Part of the challenge encountered by the Hippocratics was the care they had to show not to offend the religious predispositions of the community. As Chang claims, the Hippocratics had to walk a fine line between innovating upon current medical practices and respecting "religious beliefs of traditional Greeks that related to temple and sanctuary worship."[7] To the extent that the Hippocratics were proto-scientific and rational physicians, they were equally implicated in and committed to the religious conventions of their communities, and their writings reflected, in many places, gestures aimed at the gods. For example, the Hippocratic *Oath* begins: "I swear by Apollo Physician, by Asclepius, by Health, by Panacea and by all the gods and goddesses, making them my witnesses;"[8] and it ends with a statement on preserving and protecting "holy secrets."[9] Similarly, the Hippocratic work *Law* claims in reference to learning medicine that "[t]hings however that are holy are revealed only to men who are holy. The profane may not learn them until they have been initiated into the mysteries of sciences."[10] Even if their approach can be considered rational, it still had to be advanced on a complicated and nuanced network of differences from and similarities to the magico-religious healers.

Though a complex picture Chang presents of the relationship between the Hippocratics and their competitors, Lloyd argues for a far greater challenge. More than being rivals, Lloyd says, Hippocratic physicians and magico-religious healers shared a great deal in common with each others' methods and procedures, which shows

> just how complex the interrelations between several traditions of Greek medicine were. First some of the procedures used span both Hippocratic naturalism and temple medicine. Dreams are diagnostic tools in both, prognosis is practiced in both: the god prescribes foods and drugs and practices surgical inter-

ventions (in dreams) just as ordinary doctors did. But secondly there are cov-
ered or overt criticism from within one tradition of other rival ones. The god
warns against cautery. *On the Sacred Disease* attacks any idea of personal divine intervention in the causes or cures of diseases. Yet thirdly, the vocabulary
used to describe medical practice shows many common features. In particular
*katharsis* does double service in Hippocratic naturalism and healing that invoked the divine. It is as if both those traditions expected patients to respond
positively to the idea that what they needed was a 'cleansing.' Yet the kind of
cleansing they received proved to be very different in the two cases, for the
Hippocratics used laxatives and emetics, not charms, spells, and incantations. If
the common vocabulary suggests certain assumptions widely shared by healers
and patients, the divergent meaning and uses of the terms in question point to
the struggle to appropriate the concepts in question and give them the interpretation favored by one tradition against its rivals.[11]

Despite that some of their methods and procedures were similar, their respective statuses were nevertheless very different. Therefore, another major
challenge the Hippocratics encountered involved their perceived low status in
society, a perception formed on the basis of their unconventional practices. Instead of appealing merely to divine forces for healing, the Hippocratics were
pragmatic in their medicine, which made their practices appear as dirty and vulgar forms of manual labor. As one Hippocratic author announces, "the medical
man sees terrible sites, touches unpleasant things, and the misfortunes of others
bring a harvest of sorrows that are peculiarly his."[12] Hippocratic physicians
commonly, as part of their diagnostic protocol, handled and tasted bodily dis-
charges—including phlegm, sweat, tears, urine (and even stools!)—and palpated
their patient's inflammations, swellings, tremors, and sores. Even more alarming
is that doctors were in constant and direct contact with sickness and pollution,
witnessing and treating diseases, and becoming physically vulnerable and susceptible to them in the process.[13] As a sect of newcomers to the medical community, then, Hippocratic physicians were automatically considered of a lesser
breed, socially and professionally, because some of their ideas contradicted the
time-honored traditions of healing. The manual aspects of their methods and the
vulgar nature of their practices lowered their social standing even further, to the
point at which the author of the *Regimen in Acute Diseases* could feel comfortable in asserting—presumably among his medical peers—that the "art as a
whole has a very bad name among laymen, so that there is thought to be no art at
all."

In part due to the manual and vulgar aspects of their practices, and in part to
their more secular and pragmatic attitudes, Hippocratic physicians were regarded by the Greeks as craftsmen who sold their services in the medical marketplace to anyone who could afford them. Several ancient accounts affirm that
Hippocratic physicians were grouped together in the same social class as cob-
blers and smiths because their work required the use of their hands much more
than the application of their mind. Indeed, they were craftsmen, skilled at manual practices in medicine—cutting, stitching, bandaging—and even they regarded themselves this way.[14] The author of the *Oath*, on the other hand, attempts to distance medicine from the crude and manual practices associated with

it—to set medicine on a more intellectual trajectory by proscribing surgery and leaving that to those who are "craftsmen."[15] Not only were their practices manual, though, but they were also indiscreet about treating, as the author of the *Oath* commands, "man or women, bond or free," i.e., socially marginalized groups, like females—who were considered "polluted"—and slaves—who were often regarded as both the harbingers and the scapegoats of disease.[16]

Finally, the biggest challenge the Hippocratics encountered rose from the emergence of many fake physicians or charlatans following the spread of literacy and the dissemination of medical knowledge in the increasingly literate society of fourth century BCE Greece. As printed medical information became more readily available to a lay readership, one could profess medical knowledge—fooling people to think he were a doctor—when all he knew was what he had read in a book, with no experience in or exposure to actual medical cases.[17] Lesley Dean-Jones points out that, "it was possible for individuals to assume the status of *iatros* purely on the basis of rhetorical skill because there were no institutionally recognized credentials that marked off individuals as bona fide doctors."[18] In effect, Dean-Jones continues, it was inevitable that the Greeks were suspicious of Hippocratic physicians and could often mistake them for charlatans, or vice versa. With neither a formal or standardized system of education for doctors, nor any medical degrees or licenses to prove one's qualifications and credentials, any charlatan could feign membership in the medical community and use rhetorical skills to advertise himself as a physician. Knowing this as a possibility and having often experienced it, the public became increasingly more skeptical of doctors. Clearly, unless the doctor had a well-established reputation in the community, he had to rely on his rhetorical abilities to convince people of his technical skills and the efficacy of his treatments.

# Defending the Art of Medicine

Open debates were staged and polemics written to defend and advance Hippocratic medicine and to secure the ethos of its philosophy and its practitioners. *On the Sacred Disease*, a polemical text in the *Hippocratic Corpus* that discounts traditional approaches to treating "the sacred disease" (most likely epilepsy), is one example among many that show how the Hippocratics used the public debate forum to advance their position. Joan Leach agrees that

> The Hippocratic corpus is riddled with allusions to open debates held on medical topics in the agora. Also there are numerous allusions to and in fact, entire epideixeis treating debates over theoretical issues in the art.[19]

Hippocratic physicians participated in public debates in order to shed a positive light on their art as well as to alter the social devaluation of their practices. Following the practices of other arts, they delivered public lectures that addressed the community's concerns about Hippocratic medicine and that argued for its respective place among the arts. The author of the Hippocratic work *On the Nature of Man* is disappointed with the current state of medical debates and suggests that the best way to determine the ignorance of some doctors, especially those who propose universal theories,

> is to be present at their debates. Given the same debates and the same audience, the same man never wins in the discussion three times in succession, but now one is victor, now another, now he who happens to have the most glib tongue in the face of the crowd.[20]

*On the Nature of Man* is not alone in alluding to medical debates in the community. Many texts in the corpus give the impression that they were written as public lectures to be delivered to a general public. Living in a society that placed enormous value on speaking well, Hippocratic physicians took part in the rhetorical culture of their time, debating everything from how to heal a broken leg to the status of medicine as an art.

> Just as the sophists were instrumental in bringing about the agonistic environment that prompted other fifth and fourth century philosophers to define their terms and to structure their argument with greater precision, and just as the Second Sophistic provided Galen with rhetorical models for his public demonstrations of dissections, and spurred him to define medical terms with greater care, so some secular healers of the classical period adopted and adapted the rhetorical techniques of contemporary sophists in order to meet the expectations of audiences grown accustomed to skillfully made arguments and more convincing demonstrations of proof.[21]

The public lectures of the Hippocratics functioned in similar ways to those delivered by Sophists. As *epideixeis*, intellectual self-displays, they were a means of advertising one's expertise, generating business, and attracting disciples. Although the best among physicians might have secured an appointment—on a contractual basis—as a state-employed physician, these jobs were rare and most physicians were, instead, entrepreneurial practitioners. Even if one was lucky enough to score a state-appointed post, he had to rely on his rhetorical abilities to get it.[22] In effect, rhetoric became for the Hippocratics a technology to use in place of any formalized credentials or licensing boards that could attest to a physician's competence and skills. This is why H. F. J. Horstmanshoff goes as far as to suggest (and possibly jest) that "medicine enjoyed less prestige as a craft than it did as a subset of rhetoric."[23]

The Hippocratics, then, sought to raise themselves up in society, to be perceived as more than merely vulgar laborers, and to carve out a niche for themselves and their practices in the community. They were, as Chang describes them, "ambitious social climbers" who did what they could—socially, politically, and rhetorically—to advance their social standing and become respected and accepted practitioners in the community. Horstmanshoff thinks that "physicians who had higher aspirations put on the appearance of intellectuals, i.e., they took on the guise of rhetors in order to obscure the manual aspects of their discipline."[24]

# The Rhetorical Crafting of Identity

Through rhetoric the Hippocratics sought to create a professional identity that would be respected by the community. Evidence from the *Hippocratic Corpus* suggests that they did so deliberately and self-consciously. As we have seen, disciplinary questions about the nature of art, its standards, and its procedures, were a popular topic for the intellectuals of the time. To be recognized as part of the intellectual elite, the Hippocratics had to address the attacks that intellectuals as well as the lay public customarily levied against their art.

Some of the texts in the *Hippocratic Corpus* defend against general attacks on the status of medicine as an art. Responsive to these attacks, the author of the Hippocratic work *Law* attributes medicine's poor reputation to poor practitioners and careless critics. He claims that medicine is "the most distinguished of all the arts," but that it is now "of all the arts by far least esteemed."

> The chief reason for this error seems to me to be this: medicine is the only art which our states have made subject to no penalty save that of dishonour, and dishonour does not wound those who are compacted of it. Such men in fact are very like the supernumeraries in tragedies. Just as these have the appearance, dress and mask of an actor without being actors, so too with physicians; many are physicians by repute, very few are such in reality.[25]

The author of *The Art* cares only about defending the art of medicine, even though there are those "who have made an art of vilifying the arts."[26]

> Now as for the attacks of this kind that are made on the other arts, let them be repelled by those who care to do so and can, and with regard to those points about which they care; the present discussion will oppose those who thus invade the art of medicine, and it is emboldened by the nature of those it blames, well equipped through the art it defends, and powerful through the wisdom in which it has been educated.[27]

The same author goes on to address more specific charges that physicians only attend to "cases which would cure themselves" and that they "do not touch those where great help is necessary." This accusation is made to prove that medicine is not an art, since many people get better without the help of a doctor; others who need the most help are turned away by doctors who claim that their medicine is powerless against a certain illness.

> Some too there are who blame medicine because of those who refuse to undertake desperate cases, and say that while physicians undertake cases which would cure themselves, they do not touch those where great help is necessary; whereas, if the art existed, it ought to cure all alike. Now if those who make such statements charged physicians with neglecting them, the makers of the statements, on the ground that they are delirious, they would bring a more plausible charge than the one they do bring. For if a man demand from an art a power over what does not belong to the art, or from nature a power over what does not belong to nature, his ignorance is more allied to madness than to lack

of knowledge. For in cases where we may have the mastery through the means afforded by a natural constitution or by an art, there we may be craftsmen, but nowhere else. Whenever therefore a man suffers from an ill which is too strong for the means at the disposal of medicine, he surely must not even expect that it can be overcome by medicine.[28]

Even as in the *Laws*, medicine is extolled as the most magnificent of all arts, but it is defended in *The Art* through a frank recognition of its limits as well.

Showing his dexterity with approaching the same issue from another angle, the same author defends medicine against the charge that it is unnecessary and that people can after all contribute to their own cure.

Now my opponent will object that in the past many, even without calling in a physician, have been cured of their sickness, and I agree that he is right. But I hold that it is possible to profit by the art of medicine even without calling in a physician, not indeed so as to know what is correct medical treatment and what is incorrect, but so as by chance to employ in self-treatment the same means as would have been employed had a physician actually been called in. And it is surely strong proof of the existence of the art, that it both exists and is power-ful, if it is obvious that even those who do not believe in it recover through it. For even those who, without calling in a physician, recovered from a sickness must perforce know that their recovery was due to doing something or to not doing something; it was caused in fact by fasting or by abundant diet, by excess of drink or by abstinence therefrom, by bathing or by refraining therefrom, by violent exercise or by rest, by sleep or by keeping awake, or by using a combi-nation of all these things.[29]

Defending against these criticisms leads the author to elaborate on the art of medicine and to establish the efficacy of its healing power.

The following two passages from *The Art* demonstrate the author's sophis-ticated grasp of argument and his ability to offer a rational definition and expla-nation of an art, in the way that intellectuals are expected to, as well as how to situate its beginnings into a narrative of progress.

I will now turn to medicine, the subject of the present treatise, and set forth the exposition of it. First I will define what I conceive medicine to be. In general terms, it is to do away with the sufferings of the sick, to lessen the violence of their diseases, and to refuse to treat those who are overmastered by their diseas-es, realizing that in such cases medicine is powerless. That medicine fulfils the-se conditions, and is able constantly to fulfill them, will be the subject of my treatise from this point. In the exposition of the art I shall at the same time re-fute the arguments of those who think to shame it, and I shall do so just in those points where severally they believe they achieve some success.[30]
For the art of medicine would never have been discovered to begin with, nor would any medical research have been conducted—for there would have been no need for medicine—if sick men had profited by the same mode of living and regimen as the food, drink, and mode of living of men in health, and if there had been no other things for the sick better than these. But the fact is that sheer necessity has caused men to seek and to find medicine, because sick men did not, and do not, profit by the same regimen as do men in health.[31]

By following intellectual methods for defining, explaining, and defending an art, the Hippocratics presented themselves as intellectuals. Furthermore, they presented themselves as a distinct group tied around arguments that could be recognized as rational and argumentative procedures that were typical within the intellectual elite. Importantly, the Hippocratics advanced their arguments as responses to criticism by the public, making them appear responsive and accountable to the community. Significantly, they performed through rational argumentation their own rational approaches to medicine as a way of distinguishing themselves from traditional healers.

The Hippocratics also sought to create an identity that would alter the public's perception, by presenting their behavior as adhering to communally sanctioned norms of conduct. They employed rhetoric as part of an effort to forge themselves into a particular social class bound together by specific ways they would approach, talk to, and relate to their patients. They placed an emphasis on managing perceptions and sought to set standards for how a physician should behave and conduct himself in public. In effect, they created a rhetoric of decorum.

Across and within several of the works in the corpus, namely, *Decorum*, *Precepts*, *Oath*, *Physician*, and *Law*, medical etiquette is a prominent theme that conveys the Hippocratics' preoccupation with appearances and managing perceptions. The strategy throughout is to make medicine appear more professional in the eyes of the community. For example, one Hippocratic work advises the doctor on attending to a patient not to argue with him or with other attending physicians. *Decorum* counsels doctors on their "manner of sitting" and "bedside manner." Upon entering a patient's room, the author cautions the doctor to:

> bear in mind your manner of sitting, reserve, arrangement of dress, decisive utterance, brevity of speech, composure, bedside manners, care, replies to objections, calm self-control to meet the troubles that occur, rebuke disturbance, readiness to do what has to be done.[32]

*The Physician* advises fellow doctors to look healthy and clean, smell good, and dress well;[33] *Precepts* to avoid "luxurious headgear and elaborate perfume;" and the *Oath*, the quintessential statement on medical etiquette, proscribes assisting in abortions and engaging in sexual relations with patients, to name just a few. In effect, these texts ask physicians to pay attention to the way that they are perceived, to be conscious of their public persona. It is common to see this practical advice written into several of the texts in the *Hippocratic Corpus*.

A physician must also monitor his demeanor, making sure that it falls in line with the moral precepts of the community. His conduct as a physician should approximate that of a model member of the community. The *Physician* describes the doctor as a "prudent" person who should maintain a healthy balance in life, a gentlemanly character—marked by kindness and fairness to all—and who should exhibit self-control in dealing with "possessions very precious" like "women" and "maidens."

The prudent man must also be careful of certain moral considerations—not only to be silent, but also of a great regularity in life, since thereby his reputation
will be greatly enhanced; he must be a gentleman in character, and being this
he must be grave and kind to all . . .In appearance let him be of a serious but
not harsh countenance; for harshness is taken to mean arrogance and unkind-
ness, while a man of uncontrolled laughter and excessive gaiety is considered
vulgar, and vulgarity especially must be avoided. In every social relation he
will be fair, for fairness must be of great service. The intimacy also between
physician and patient is close. Patients in tact put themselves into the hands of
their physician, and at every moment he meets women, maidens and possessions very precious indeed. So toward all these self-control must be used. Such
then should the physician be, both in body and in soul.[34]

Clearly, the rhetoric of decorum delineated in the corpus extends to all aspects
of public and private life.

Another aspect of etiquette addressed concerns the communication dynam-
ics between a physician and his patient. *Decorum* advises the physician to speak
cautiously and to avoid discussing with a patient upsetting and stressful news
that could affect their recovery.

Perform all this calmly and adroitly, concealing most things from the patient
while you are attending to him. Give necessary orders with cheerfulness and se-
renity, turning his attention away from what is being done to him; sometimes
reprove sharply and emphatically, and sometimes comfort with solicitude and
attention, revealing nothing of the patient's future or present condition. For
many patients through this cause have taken a turn for the worse, I mean by the
declaration I have mentioned of what is present, or by a forecast of what is to
come.[35]

*Precepts* similarly warns that a patient's recovery is inexplicably linked to discussing matters that make a patient uncomfortable—like the physician's fee.
The author pleads with the doctor to be fair, and to consider not only his patient's financial situation but his physical condition as well.

I urge you not to be too unkind, but to consider carefully your patient's supera-
bundance or means. Sometimes give your services for nothing, calling to mind
a previous benefaction or present satisfaction. And if there be an opportunity of
serving one who is a stranger in financial straits, give full assistance as such.
For where there is love of man, there is also love of the art. For some patients,
though conscious that their condition is perilous, recover their health simply
through their contentment with the goodness of the physician. And it is well to
superintend the sick to make them well, to care for the healthy to keep them
well, but also to care for one's own self, so as to observe what is seemly.[36]

Physicians were expected to respect their patients and be sensitive to the psychological and physiological effect that their "goodness" can evoke.

> For should you begin by discussing fees, you will suggest to the patient either that you will go away and leave him if no agreement be reached, or that you will neglect him and not prescribe any immediate treatment. So one must not be anxious about fixing a fee. For I consider such a worry to be harmful to a troubled patient, particularly if the disease be acute. For the quickness of death, offering no opportunity for turning back, spurs on the good physician not to seek his profit but rather to lay hold on reputation. Therefore it is better to reproach a patient you have saved than to extort money from those who are at death's door.[37]

The Hippocratic *Oath* is definitive with matters of conduct and etiquette and it may have been part of a doctor's initiation into the Hippocratic "guild." This oath was likely recited by a physician in front of a community of his peers, toward the end of a long apprenticeship with a more experienced and venerable doctor. Even to this day, most medical schools require graduating doctors to recite the *Oath* and pledge allegiance to the medical code of conduct, although the oath they recite is a modernized version of the Hippocratic one.

Finally, the Hippocratics sought to create an identity that would display their expertise demonstrably to their patients, and would enable the public to differentiate them more readily from impostors; they were able to do this through the practice of prognosis. As the author of *Prognosis* declares, this practice allows the physician to fill in the blanks in the patient's fragmented narrative recounting of his illness.

> I hold it is an excellent thing for a physician to practice forecasting. For if he discover and declare unaided by the side of his patients the present, the past and the future, and fill in the gaps in the account given by the sick, he will be the more believed to understand the cases, so that men will confidently entrust themselves to him for treatment.[38]

Prognosis affords the physician with a rhetorical tool for securing his patient's trust and confidence in his methods and treatments, by predicting on the basis of past experiences the future course of his patient's illness. Prognosis also taps into the cultural logic of the times which places a great deal of respect and authority in prophecy and forecasting; so much so that the author of *Decorum* cautions that forecasting should begin before entering the patient's room.

> When you enter a sick man's room, having made these arrangements, that you may not be at a loss, and having everything in order for what is to be done, know what you must do before going in. For many cases need, not reasoning, but practical help. So you must from your experience forecast what the issue will be. To do so adds to one's reputation, and the learning thereof is easy.[39]

Prognosis was a valuable rhetorical tool that the Hippocratics used to gain their patient's trust, confidence, and compliance, as well as to endow their art with a culturally sanctioned authority. In Lloyd's words,

> Unless he already possessed an established reputation, the itinerant doctor was faced with a recurrent problem in having to build up a clientele in each city he stayed in. Here the practice of 'prognosis', which included not only foretelling the outcome of a disease, but also describing its past history, was an important psychological weapon.[40]

Prognosis was the ultimate way for the itinerant physician to make a respected claim about his expertise and to establish through it the reputation in the community that he hoped for. By means of causal narratives that gave to a patient's illness an ordered continuum of a past, present, and future, the Hippocratics used the art of narration to constitute their identity as experts.

# Conclusion

This chapter has aimed at shedding light on the rhetorical history of medicine by examining the ways through which the Hippocratics employed rhetoric to attain their goals as a group, thereby securing their respective position in the history of Western medicine. Their goals were connected to the drive to overcome the obstacles confronting them from the start: the competition they faced from well-established magico-religious healers; the low-class status as craftsmen attached to them due to their novel and pragmatic approaches to treatment; and the ill-reputation that plagued them due to the rising number of impostors who mimicked their trade. Rhetoric provided an opportunity for the Hippocratics to meet their goals by overcoming these obstacles—sometimes through public demonstrations and defenses of their art, other times through laudable deeds and displays of professional etiquette and conduct.

In rhetoric, the Hippocratics found ways to craft an identity as a distinct group of healers, an identity they could display publicly to exhibit their uniqueness as a group and to consolidate their methods into an identifiable set of common practices. Like the Sophists, they used the art of argumentation to display this identity to the intellectual elite and to defend the legitimacy of their practices through reasoned discourses that followed established procedures for defining, explaining, and defending an art in public. Like other social climbers, they forged their identity around norms of conduct that extended from appearance and etiquette, through interactions with patients, to general patterns of behavior as members of the community. Like tellers of stories, they used their skill with narratives to display their prognosis of illnesses to the public, offering coherent accounts of illnesses and symptoms that tied the present with the past and the future, and strengthening their credibility as experts while enhancing their distance from impostors. Thus crafting their identity rhetorically, the Hippocratics demonstrated their dexterity to appropriate rhetoric for their own concerns.

# Notes

This paper was initially delivered in 2009 at the Twenty First International Conference on Philosophy in Paphos, Cyprus, sponsored by the International Association for Greek Philosophy and the International Center for Greek Philosophy and Culture. The author thanks the conference participants for their insightful suggestions for revision.

1. When I use "Hippocratics" I am referring to the disciples of Hippocrates who worked and studied with him or who later followed his methods and practices in the fifth and fourth centuries.

2. The *Hippocratic Corpus* refers to the body of works posthumously attributed to Hippocrates. In fact, all of the texts are anonymous, and there is little possibility that the historical Hippocrates authored any of them.

3. G. E. R. Lloyd, *In the Grip of Disease: Studies in the Greek Imagination* (New York: Oxford University Press, 2003), 40.

4. The *Oath* is one example.

5. Julie Laskaris, *The Art is Long: On the Sacred Disease and the Scientific Tradition* (Leiden: Brill, 2002), 31.

6. Hui-hua Chang, *Testing the Serpent of Asclepius: The Social Mobility of Greek Physicians* (PhD diss, Indiana University, 2003), 152.

7. Hui-hua Chang, Testing *The Serpent of Asclepius,* 45.

8. *Oath,* 1

9. *Oath,* 30.

10. *Law,* V

11. Lloyd, *In the Grip of Disease,* 58.

12. *Breaths,* 2.227.

13. Thucydides remarks in his telling of the Peloponnesian War that when the plague hit Athens toward the beginning of the War, "the doctors were quite incapable of treating the disease because of their ignorance of the right methods. In fact, mortality among the doctors was the highest of all, since they came more frequently in contact with the sick." Thucydides, *History of the Peloponnesian War.* Penguin Books 150: 47 (1954).

14. *On Ancient Medicine,* 1, 7-12.

15. The section from the *Oath* reads: "I will not use the knife, not even, verily, on sufferers from stone, but I will give place to such as are craftsmen therein" (20-21).

16. *Oath,* 27.

17. In *Phaedrus* (268a-c), Plato calls this type of person "crazy" if he professed to be a doctor when all he knew about medicine and administering drugs was from what he had read in a book rather than from real medical practice and experience. It should be remembered that Greece in the fifth century is a primarily oral culture, and medicine was an art passed down orally from one generation to the next. Therefore, the move toward writing and recording medical information for recall or pedagogical purposes marks an important deviation from accepted norms in the medical community.

18. Lesley Dean-Jones, "Literacy and the Charlatan in Ancient Greek Medicine," in *Written Texts and the Rise of Literate Culture in Ancient Greece,* ed. Harvey Yunis (Cambridge: Cambridge University Press, 2003), 97.

19. Joan Leach, *Healing and the Word: Hippocratic Medicine and Sophistical Rhetoric in Classical Antiquity* (PhD diss, University of Pittsburgh, 1996), 62.

20. *On the Nature of Man,* 20-28.

21. Laskaris, 76.

22. Ludwig Edelstein argued that the only really promising career objective for a doctor was to become a state-employed physician. See Edelstein, *Ancient Medicine: Collected Papers of Ludwig Edelstein*, ed. Owsei Temkin and C. Lillian (Baltimore: Johns Hopkins University Press, 1969).

23. H. F. J. Horstmanshoff, "The Ancient Physician: Craftsman or Scientist?" *Journal of the History of Medicine* 45 (1990): 185.

24. Horstmanshoff, 193.

25. Horstmanshoff, 1-8.

26. *The Art*, I, 1-4.

27. *The Art*, I, 18-23.

28. *The Art*, VIII, 1-20.

29. *The Art*, V, 1-18.

30. *The Art*, III, 3-15.

31. *On Ancient Medicine*, III, 1-12.

32. *Decorum*. XII, 1-6.

33. *Physician*, 1-7.

34. *Physician,* 8-29.

35. *Decorum*, XVI, 1-10.

36. *Precepts*, VI.

37. *Precepts*, IV, 8-14.

38. *Precepts*, IV. 1-8.

39. *Decorum*, XI, 1-7.

40. G. E. R. Lloyd, *Early Greek Science: Thales to Aristotle* (New York: W.W. Norton & Company, 1970), 52.

# Chapter 2

# The Wisdom in Wood Rot: Finding God in Early-Modern Scientific Explanation

Eric Palmer

This chapter presents a historical study of how science has developed and of how philosophical theories of many sorts—philosophy of science, theory of the understanding, and philosophical theology—both enable and constrain certain lines of development in scientific practice. Its topic is change in the legitimacy or acceptability of scientific explanation that invokes purposes, or ends; specifically in the argument from design, around the turn of the eighteenth century.

We tend to explain various features of the world by invoking purposes. Philosophers of science and scientists also tend to accept this kind of explanation as legitimate, as is most obvious in our time in, for example, archaeology. It does not stretch credibility to suggest that the specific orientation of a building footprint or the pattern of elements in assemblages of monumental architecture arose due to the builders' regard for the alignment of the sun at noteworthy times, such as briefest and lengthiest days of the year, the days on which night and day are of equal length (the vernal and autumnal equinox), or the beginning of a planting season. A comprehensive explanation for why Mayan temples and cities are just as they are will include reference to the intentions of their designers. Of course, the same holds for buildings through time, the world over.[1]

But what are we to make of an introduction of purposes such as the following, which was provided by a prominent French author of the mid-eighteenth century?

> [W]here is the Goodness, it may be objected, in [God] having created . . . destructive Worms, for Example, which insensibly eat and consume the Sides of our Ships, the piles of our Dikes, and the Timber of our Houses?
> These Worms, like all others, do by the Corruption of one thing contribute to the Generation of another, and serve to promote the general Circulation of the Commodities and Productions of different Countries, on which Commerce necessarily depends. So mean an Animal, in Appearance, as the Pipe-Worm, by usefully employing the vigilance of the Dutch, not only maintains but brings Riches to the inhabitants of Sweden. . . . Were they not under a perpetual Necessity of tarring and sometimes repairing their Vessels and Dikes at Amsterdam, in vain would the Muscovite and Norwegian barrel up the Pitch, which

distils from their Pines; in vain would the Swedes cut down the Oaks and Lofty
Fir-Trees that grow in their Forests. Thus does this little Animal, which we so
much complain of as troublesome and injurious to us, become the very Cement,
which unites these distant Nations in one common Interest. . . . The Prospect
we have taken of Nature, does in every part sufficiently prove that the Good of
Man was the chief End proposed by Providence in the Works of the Creation,
even in those very things, which seem hurtful or offensive.[2]

The explanation invokes purposes, much like the human purposes found in the
archaeological explanation noted above. Yet we would consider the two cases
very different: the wood rot explanation invokes God's purposes, and in that respect, it is explanation of a very different sort than we expect to find within current archaeology and social sciences generally.

The purpose for wood rot suggested here might move a twenty-first century
audience to laughter. It appears to be much the sort of claim that Voltaire's Dr.
Pangloss would make within the pages of *Candide*. Compare, for example, the
Panglossian explanation of the arrival of syphilis in Europe following contact
with the New World:

> It was a thing unavoidable, a necessary ingredient, in the best of worlds! For
> if Columbus had not landed upon an island in America, and there catched
> this disease—which contaminates the source of life, frequently hinders gen-
> eration, and is evidently opposite to the great end of nature—we should have
> neither chocolate nor cochineal.[3]

Indeed, the similarity is not accidental: Voltaire was ridiculing this sort of explanation, and the very author of the discussion, the abbé Noël Pluche, in particular.[4] Pangloss provides his disquisition for the education and spiritual edifica-
tion of a promising and inquisitive youth. Pluche had a similar end: his example
of wood rot appears in a remarkably popular publication, a multi-volume com-
pendium of science, human invention and piety entitled, in its English translation, *Nature display'd: Being discourses on such particulars of natural history
as were thought most proper to excite the curiosity and form the minds of youth.*
Though the above explanations pertain to "goodness" and "the best of
worlds" respectively, they both have very much to do with detailed accounting
of natural and human history. Both educators draw our attention to the ubiquity,
even within literary culture, of a little-studied aspect of early modern scientific
explanation: the rise of references to God's purposes within the study of nature,
starting in the second half of the seventeenth century. The most influential theorists of scientific knowledge shortly before that time, Francis Bacon and René
Descartes, would summarily ban such explanation from natural philosophy.
References to divine purposes and divine beneficence would gain philosophical
credentials—for the first time in the context of modern science—in English natural philosophy, under the influence of prominent scientists and philosophers
who were members of the Royal Society of London for the Improving of Natural Knowledge (henceforth Royal Society), particularly in the final quarter of the
century. Robert Boyle would produce the major philosophical justification for
such explanation and significant roles promoting it would be played by a net-

work of members, including Isaac Newton. From those beginnings, the form of explanation that soon came to be known as physico-theology would become a well-considered topic for theologians and popularizers of science such as Pluche and would gain respectability as a pursuit for practicing scientists over the course of two hundred years.

Naturalists from Aristotle on have made reference to purposes in their treatment of plants and animals, including both goals toward which apparently mindless activity is directed and the uses or the functions of the parts of organisms. Goals and uses would appear to be the product of intention and design respectively, and so of a mind or a designer. But discussion of that designer grew rapidly late in the seventeenth century, and was pursued by scientists as well as clergy and popularizers of science. Pluche's surprising thesis was not far off of what a respected scientist might write. His contemporary Carl von Linné (Carolus Linnaeus) repeatedly wrote on the characteristics of the divine designer. In 1751 he would write, "What genius, what art, can imitate one of those fibres whose various and infinite complications form the human body? In its most minute filament we see the finger of God, and the seal of the great Artificer of Universal Nature."[5] It was his custom to work collaboratively on the dissertations of his doctoral students, and so, two years earlier, either he or Isaac Biberg wrote that "Goats . . . have feet made for jumping." The co-authors extended the purposes that pertain to animals beyond the roles of the parts of the organism, beyond activity advantageous to the organism, and beyond kin and species benefits as well. They argued for goals for activities that also appear to entirely escape the creatures' understanding and learning, in flights of analysis almost as memorable as those of Pluche:

> As the excrement of *dogs* is of so filthy and septic a nature, that no *insect* will touch them, and therefore they cannot be dispersed by that means, care is taken that these animals should exonerate upon stones, trunks of trees, or some high place, that vegetables may not be hurt by them. *Cats bury their dung.* Nothing is so mean, nothing is so little, in which the wonderful order, and wise disposition of nature does not shine forth.

By whom was "care taken?" The source of the "wise disposition of Nature" toward the preservation of vegetables was swiftly identified as divine providence intended foremost for human good. The authors conclude, "all these treasures of nature so artfully contrived, so wonderfully propagated, so providentially supported throughout her three kingdoms, seem intended by the Creator for the sake of man."[6]

Physico-theology would be pursued with reduced vigor on the continent of Europe after 1750, but it would remain strong in England and would see an impressive resurgence early in the following century. The English movement reached a second peak with a cast of British scientists including William Whewell, William Buckland and William Prout producing full volumes on cosmology, geology, chemistry, and other fields under the series title *Bridgewater Treatises on the Power, Wisdom and Goodness of God, as Manifested in the Creation* (1833-36, 8 vols.).[7] The criticism of David Hume and Immanuel Kant hardly affected the pursuit of physico-theology; Darwin ushered it off the stage during his lifetime.[8] *The Origin of Species* (1859) presents Darwin's theory of natural selection, in which, over generations, heritable natural variation and selective retention of reproducing organisms provides an alterative explanation of order that design may resemble.

# Natural theology in flux

Natural theology isn't what it used to be. Around the turn of the eighteenth century it developed a particularly close association with science that was reflected in the introduction of a new term, "physico-theology." The rise of this association is evident in authors' characterizations of the term in philosophical works from different historical periods, as I will endeavor to show below.

Natural theology—a term used interchangeably with "natural religion"—finds a brief and clear characterization that reflects a close association with natural philosophy in *Of the Principles and Duties of Natural Religion* (1675). The English cleric and polymath John Wilkins (1614-1672), very near to the end of his life, writes, "I call that *Natural Religion*, which men might know . . . by the meer principles of Reason, improved by Consideration and Experience, without the help of *Revelation*." Wilkins' treatment presents a balance between reason and experience: the former is contrasted with and aided by the latter in the production of knowledge. This is an approach we might expect of a European philosopher after the middle of the seventeenth century, in a field shaped especially by the discussions of Descartes and Hobbes. Wilkins, who does not himself "pretend to the invention of any new arguments," launches into "the most plain and convincing" independent lines of natural theological argument. He finds the best basis for knowledge of God in: "(1) The Universal consent of Nations, in all places and times. (2) The Original of the World. (3) That excellent contrivance which there is in all natural things. (4) The works of Providence in the Government of the World."[9] In Wilkins' first three topics, the ground is prepared for the connection of natural religion to areas of study that would later be classified as sociology, cosmology, physics and biology.

Natural theology is much older, however, and its past is very different. Wilkins' treatment tacitly dismisses the greater proportion of what would have been central to its study by all previous generations of philosophers. Consider, for comparison, the difference in emphasis found in two other treatments, one from long before by Varro Reatinus (116-27 BCE) and one from early in Wilkins' own century by Francis Bacon (1561-1626).

Varro's text survives through quotations contained within the writing of Augustine (354-430 CE). Varro characterizes natural theology as:

> that concerning which philosophers have left many books, in which they treat such questions as these: what gods there are, where they are, of what kind and character they are, since what time they have existed, or if they have existed from eternity; whether they are of fire, as Heraclitus believes; or of number, as Pythagoras; or of atoms, as Epicurus says; and other such things, which men's

ears can more easily hear inside the walls of a school than outside in the Forum.[10]

Polytheism is a going concern here, but the relation of reason to experience is not noted. Different times reflect different philosophical concerns: Varro's reference to eternity evokes one topic of philosophical theology that is common in the history of philosophy, which we know as the cosmological argument. Aquinas' familiar arguments for the existence of God, the Five Ways, include two formulations of the argument: one concerns the necessary existence of an unmoved mover, the second argues for a first cause that is necessary for all that follows. Natural theology isn't what it used to be: though each author might find some of the arguments noted by the other to be agreeable, there is no overlap among Wilkins' preferences and Varro's references.[11]

Aquinas does present some overlap with Wilkins in his Five Ways. The fifth way, in its entirety, follows:

> The fifth way is taken from the governance of the world. We see that things which lack intelligence, such as natural bodies, act for an end, and this is evident from their acting always, or nearly always, in the same way, so as to obtain the best result. Hence it is plain that not fortuitously, but designedly, do they achieve their end. Now whatever lacks intelligence cannot move towards an end, unless it be directed by some being endowed with knowledge and intelligence; as the arrow is shot to its mark by the archer. Therefore some intelligent being exists by whom all natural things are directed to their end; and this being we call God.[12]

The argument is among the ancestors of physico-theology. Traces of a related argument were voiced by Socrates, as reported by Xenophon. Socrates follows a discussion of the utility of eyebrows, eyelids and eyelashes with two queries:

> Does it not strike you then that he who made man from the beginning did for some useful end furnish him with his several senses—giving him eyes to behold the visible word, and ears to catch the intonations of sound? . . . I ask you, when you see all these things constructed with such show of foresight can you doubt whether they are products of chance or intelligence?[13]

Socrates' move shows a specific similarity to the wood rot argument, and more particularly, to physico-theological arguments that concern the functions of parts of animals, like those of Linnaeus and Biberg. These arguments differ from the cosmological argument in that, though the cosmological argument refers to the observed universe, it does not refer to a divine "end" or "foresight" to be found in the production of a specific feature within the universe. Arguments of this sort have since become known as teleological arguments for the existence of God, or, in a phrase, "argument from design": argument from a survey of the design of the world or a portion of the world to a conclusion concerning the existence or characteristics of the designed object's maker. It is evident from Xenophon's writing that teleological argument is longstanding in philosophy.

Like Socrates and Aquinas, Wilkins and other moderns who will be considered below survey observable aspects of the world to support argument to establish that God exists. But such proof—probable rather than necessary, or, in Wilkins' terminology, "morally certain," rather than "mathematically certain"—the moderns consider very easy to achieve. The arguments of physico-theology are greatly expanded in detail and altered in purpose: beyond using empirical considerations to demonstrate God's existence, the target of physico-theology is proof of the "Power, Wisdom and Goodness of God," as the series title of the *Bridgewater Treatises* suggests. Providence, which is just hinted at by Socrates, is the focus of investigation.

Wilkins' text is an early indicator of the consolidation of an intellectual shift comprised of two complementary aspects. One aspect is the rise of a modern theological sensibility regarding to nature, a sensibility that develops as a consequence of a diminution of the symbolic significance of nature that is evident in Renaissance humanism. The second and subsequent aspect is the rise of a modern scientific and philosophical sensibility concerning the divine, a change that comes as a consequence of the development of both empiricism and the new theory of the understanding. Modern philosophy changes the game, redefining the possibilities for natural theology and ushering in the new physico-theology after theology has itself altered to become more consonant with the ideals that would be expressed in modern empiricism. The first aspect of the double shift is already evident in the work of Francis Bacon; both faces appear in Wilkins.

Concerning the first aspect, the development of a theological perspective that is modern in some respects, consider the narrow band of experience that is relevant to natural theology as indicated (in italics I have inserted) in our third characterization of natural theology, from Bacon's *Advancement of Learning* (1605):

> as concerning divine philosophy or natural theology, it is that knowledge or rudiment of knowledge concerning God, which may be obtained *by the contemplation of His creatures*; which knowledge may be truly termed divine in respect of the object, and natural in respect of the light. [14]

This indicates a narrowing of the topic, as is evident also in Wilkins, such that natural theology no longer contains all of philosophical theology: it is no longer a topic that stands in straightforward contrast to revealed religion. Philosophical theology such as Anselm's ontological argument for the existence of God would not actually fall within Bacon's characterization, for example. Arguments that focus upon metaphysical necessity, such as cosmological arguments, are not excluded, since contemplation of creation is rather abstractly involved in thinking about chains of causes, but such metaphysical argument does not leap to one's mind when reading this characterization, either. The narrowing directs us toward philosophical theology that highlights empirical considerations.

Bacon also shows a modern theological sensibility concerning nature, and illuminates the shift itself, within the following:

> for our Saviour saith, *You err, not knowing the Scriptures, nor the Power of*

*God*; laying before us two books or volumes to study, if we will be secured from error; first, the Scriptures, revealing the will of God; and then the creatures expressing His Power; whereof the latter is a key unto the former; not only opening our understanding to conceive the true sense of the Scriptures, by the general notions of reason and rules of speech; but chiefly opening our belief, in drawing us into a due meditation of the omnipotency of God, which is chiefly signed and engraven upon His works.[15]

This image of "two books" provided for our enlightenment by God, along with the exhortation to study both, has a long history. It becomes an authoritative rationale for the pursuit of natural philosophy as it is develops from a sketch of the various paths to knowledge of God presented in the writing of Aquinas.[16] It finds support in a variety of biblical passages and it inspired Raymond Sebond's *Theologia Naturalis* (1436), which styles the world as "composed of a great multitude of creatures, like a collection of letters in a book."[17] Lessons intended for man are written in the book of nature by God, particularly indicating "the ladder of nature that man climbs up to understand himself and his creator." Sebond's nine-hundred page treatise includes teleological argument concentrated over just a few pages, and that discussion is set within an effort to use observation of nature as an aid to understanding the unity of the natural order, with God, the most perfect being, at its top, and humanity second.[18] Sebond's natural theology, then, is not so much a reflection upon design as it is a meditation upon order, particularly as displayed in hierarchy, or the chain of being. Sebond focuses upon specific lessons concerning that order that are thoughtfully introduced into the scheme, and so provided for us, by God.

Bacon's assumptions regarding the uses of the book of nature are far more parsimonious than Sebond's, and the lessons to be learned are also less extensive. In Bacon, nature is stripped of the allegorical significance that is found in Sebond and displayed much more broadly in the vogue for an emblematic interpretation of the world that blossomed in the second quarter of the sixteenth century.[19] Bacon is a modern, rather than a Renaissance humanist: he will not argue that nature is arranged expressly for our philosophical instruction, even if he might consider such a thing plausible, for that would reflect an extravagance of hypotheses not fitting with the empirical character of natural philosophy, in which "the basis is natural history; the stage next the basis is physic; the stage next the vertical point is metaphysic."[20]

Though Bacon suggested that the study of nature would provide a "due meditation of the Omnipotency of God," his own meditation was abstract and limited by comparison with those appearing later in the century. For Bacon, natural theology sketches the "rudiment" of knowledge concerning God: "that God exists, that he governs the world, that he is supremely powerful, that he is wise and prescient, that he is good, that he is a rewarder, that he is an avenger, that he is an object of adoration—all this may be demonstrated from his works alone." Bacon limits natural theology both by erecting methodological walls and by expressing doubts. First, he promotes a division of disciplines that ensures that science and metaphysics are pursued separately. Bacon opposes the invocation of design in explanation for natural science, or "physic": he limits natural science to inquiry into material and efficient causes and he charges that the search for answers regarding purposes tends to stunt empirical study, leading thinkers to halt their inquiry into material and efficient causes. He writes:

> that the clouds are for watering of the earth; or that the solidness of the earth is for the station and mansion of living creatures; and the like, is well inquired and collected in metaphysic, but in physic they are impertinent. Nay, they are, indeed, but *remoras* and hindrances to stay and slug the ship from farther sailing, and have brought this to pass, that the search of the physical causes hath been neglected and passed in silence.[21]

Bacon does not make it apparent how one might pursue these lines of inquiry in metaphysics. He would have been skeptical of detailed natural theology of the sort that appears late in the century: he writes, "I hold it is not possible to be invented by that course of invention," and quotes Ecclesiastes 3:11: "The work which God worketh from the beginning to the end, it is not possible to be found out by man." He holds no expectation of future advance in this area: "the summary law of nature, we know not whether man's inquiry can attain unto it." In a passage concerning purposes that lays bare Bacon's views on politics as well as divine mystery, he suggests:

> For as in civil actions he is the greater and deeper politician, that can make other men the instruments of his will and ends, and yet never acquaint them with his purpose, so as they shall do it and yet not know what they do; than he that imparteth his meaning to those he employeth: so is the wisdom of God more admirable, when Nature intendeth one thing, and Providence draweth forth another; than if He had communicated to particular creatures, and motions, the characters and impressions of His Providence. [22]

Descartes would even exceed Bacon in caution concerning final causes in science. In the *Principles of Philosophy* (1641) he would argue very briefly that "We should not be so arrogant as to suppose that we can share in God's plans."[23]

## Natural theology and the Royal Society

Later in the century and generations before Pluche, English philosophers shook off such skepticism. Bacon's writing was taken to align with the empirical ideals of the new natural science developed in the Royal Society, of which Wilkins was a founding Fellow and the Secretary from inception in 1660. Indeed, Bacon was found to be prescient: Abraham Cowley's ode "*To the* Royal Society" (1667), casts Bacon's *New Atlantis* of 1623 as a work that prophesies the formation of the Royal Society. In an image that neatly ties the religious and the scientific, Cowley places Bacon as the Mosaic leader of English empirical philosophy:

Bacon, *like* Moses, *led us forth at last,*
*The barren Wilderness he past,*
*Did on the very Border stand*
*Of the blest promis'd Land,*
*And from the Mountain's Top of his exalted Wit*
*Saw it himself, and shew'd us it.*[24]

Within the promised land, the teleological argument began its expansion into the fresh fields of physico-theology. Henry Power unequivocally announced a proper wedding of natural theology and the new mechanical philosophy in 1662: "all things are Artificial; for Nature it self is nothing else but the Art of God. Then, certainly, to find the various turnings, and mysterious process of this divine Art, in the management of this great Machine of the World, must needs be the proper Office of only the Experimental and Mechanical Philosopher."[25] Robert Hooke presented similar sentiments, casting God as engineer in *Micrographia* (1665). The relatively crude shaping of a pin's point seen under magnification compared poorly to the fineness of "the *hairs*, and *bristles*, and *claws* of multitudes of *Insects*." He concluded that the microscope reveals that, in man's efforts, there is "rudeness and bungling of *Art*":

> the more we see of their *shape*, the less appearance will there be of their *beauty*: whereas in the works of *Nature*, the deepest Discoveries shew us the greatest Excellencies. An evident Argument, that he was the Author of all these things, was no other then *Omnipotent*; being able to include as great a variety of parts and contrivances in the yet smallest Discernable Point, as in those vaster bodies (which comparatively are called also Points) such as the *Earth, Sun,* or *Planets.*

Hooke extends the teleological argument over new ground here, but physico-theology is still in early development. *Micrographia* is frequently cited as a key text by figures in the late seventeenth century and recent scholars, yet it contains only fleeting references to "the Authour of all," plus several hundred words of directed physico-theological argument.[26] The new empirical science would radically reshape natural theology as it came to develop among members of the Royal Society in the final quarter of the century. Like Socrates, John Wilkins argued from the good design of the human body, concluding, "From whence it will follow, That it must be a Wise Being that is the Cause of these Wise Effects." Wilkins would develop his argument at a full chapter's length, much greater than Socrates and Hooke, citing both ancient and contemporary science: Galen on the complexity of the human body and Hooke's observations of God's craftsmanship through the microscope. [27]

The importance of the new empirical science to his effort actually appears to lead Wilkins to downgrade other approaches to natural theology, such as the cosmological argument. There is another philosophical factor, however, that reduced their value further. The cosmological argument, which before had claimed the status of demonstrative or apodictic argument, appears to be ignored or tacitly dismissed in *Natural Religion*. Wilkins' book does include discussion of a necessary existent and a first mover but his taste for such argument is greatly tempered and is clearly affected by a theory of the human understanding that is in the process of reshaping English philosophy. Wilkins does not attempt to develop a cosmological proof for God's existence, instead he writes: "The most general Notion that men have of God, is that He is the first cause, and a Being of all possible Perfection."[28] The reference to a "general notion," suggests the place of the idea of God within a developing theory of ideas that suffuses Wilkins' text and is a recognizable antecedent of the account to be found in John Locke's *Essay Concerning Human Understanding* (1690).

The second aspect of the intellectual shift noted above, concerning new philosophical trends that reshape natural theology and is displayed in Wilkins' approach. First, natural philosophy has swept Wilkins' discourse of natural theology, to such an extent that scant attention is paid to what were before considered the more secure apodictic forms of argument, such as the cosmological argument. Second, the new theory of the human understanding evidently plays a large role in undermining the claim to certainty held by those other forms of argument. The efforts in natural philosophy and theory of ideas were seen as linked by Wilkins, who explicitly proposed the improvement of language as an important task to the Royal Society in 1668, and by Locke, who implicitly did the same, referring to himself as an "under-labourer" to "master-builders" such as Newton, Boyle and Huygens, "removing some of the rubbish that lies in the way of knowledge."[29] The new theory of ideas posed a significant barrier to most traditional forms of natural theology. The proposal that the mind is a *tabula rasa* upon which experience is impressed presented an acute challenge to innate ideas, and so, to arguments for the existence of God that were not suitably grounded in experience. Consequently, many of the most prominent physico-theologicans chose to soft-pedal or entirely forego "metaphysical" argument for the existence of God, referring readers to other sources or skipping traditional apodictic argument entirely. They would gesture at such paths, indicating that little attention was required on routes well-trodden by others; or they would straightforwardly declare that they "always esteemed the strongest" approach to natural theology to be physico-theology.[30]

Locke stands as the most important proponent of the new and influential theory of ideas. Robert Boyle is the most important theorist of natural philosophy for this new philosophical turn. He would address the topic of final causes in science at length in *A Disquisition About the Final Causes of Natural Things: Wherein it is Inquir'd, Whether, And (if at all) With what cautions a Naturalist should admit Them?* (1688) Boyle's "cautions" are limited, showing only vestiges of Bacon's concerns. He opens his disquisition by commending the importance of the subject, admonishing his reader, "if we neglect this Inquiry, we live in danger of being Ungrateful, in overlooking those Uses of Things that may give us Just Cause of Admiring and Thanking the Author of them . . ." Boyle argues that knowledge of final causes is attainable through empirical inquiry: in this he opposes Epicurean mechanists and Cartesian skeptics who "suppose all the ends of God in Things Corporeal to be so Sublime, that 'twere Presumption in Man to think his Reason can extend to Discover them."[31]

Boyle takes great care to distinguish four categories of ends that may be the subject of inquiry. The first category is the purpose for the entirety of the cosmos, "Exercising and Displaying the Creators immense power and admirable Wisdom." The second category comprises large-scale systems within the cosmos: "Ends design'd in the number, fabrick, placing, and wayes of moving the great Masses of Matter, that, for their Bulks or Qualities, are considerable parts of the World . . . sun, moon, and fixed stars, and the terraqueous Globe . . ." The third covers "the Parts of Animals . . . destinated to, and for the welfare of the whole Animal himself, as he is an entire and distinct System of organiz'd parts, destinated to preserve himself and propagate his *Species* . . ." Finally, Boyle cites a fourth sort of Ends, "call'd *Human* Ends, which are those that are aim'd at by Nature, where she is said to frame Animals and Vegetables, and other of her productions, for the use of Man."[32] The scheme appears to be incomplete and arbitrary in at least this respect: the second category includes purposes pertaining to properties of and interrelations among very large bodies, but it tacitly excludes purposes that pertain to interrelations among middle-sized objects, both animate and inanimate. Such interrelations among bodies—worms and wood, dogs and vegetables, *etc.*—appear to be neglected by Boyle. I will call such collections "assemblages," regardless of whether they are large or middle-sized (and we might add "small" to the list, to complete it and find a place for chemistry). Relations characterized as assemblages are to be contrasted with the relations among parts of an organism.

Consider how Boyle might have, but did not, discuss the assemblage that produces wood rot. It would appear that the divinely designed role that burrowing worms play in breaking wood down reflects an end that might have fit within his second category. Boyle's discussion of the divine purposes for assemblages includes mention of the role of the sun in furnishing the earth with heat and light, and the relation of the "two Chief parts" of the globe, the continents and the oceans. Boyle does not descend to a smaller scale, however: from these global assemblages he jumps to discussion of the design of the parts of organisms.[33] Late in the work, Boyle explains his choice: he finds that one might discern God's "particular providence" concerning the purposes of parts of animals, whereas "it is not an easie Task" to inspect assemblages and discern the plan of "general providence." How worms are designed to eat is easy to see; how worms and ships fit together, and into the general plan of providence, is not easy to see. The best that Boyle can offer he identifies as supposition: that there is "One Grand Motive" to the whole of creation, that "might, by so many and so very differing Contrivances, as are to be met with in the Structure of Men, Four-footed Beasts . . . etc., Exercise and Display . . . *the Multifarious or Manifold Wisdom of God. (Ephesians 3.10)*"[34]

To go beyond conjecture and explain the interrelations of such assemblages, Boyle adds two supports in combination: anthropocentrism and revelation. Anthropocentrism puts such relations into the fourth category of Boyle's division, but to acquire knowledge in that division it is not sufficient for the "Naturalist to discourse merely on physical grounds." When we add "Revelations, contained in the Holy Scriptures, we may Rationally believe more, and speak less Hesitantly, of the Ends of God, than bare philosophy will warrant us to do." With the supplement of revelation, then, Boyle comes to be comfortable discussing assemblages that are similar to the pair of worms and wood. For example, he quotes the book of Genesis to support a detailed claim: "God *deliver'd all Terrestrial Beasts, and Fowle and Fishes, and Every moving thing that lives, into the hands of Men*; and intended that they should eat Animals . . ." In this context, Boyle states that the sun was meant by God to grow plants "that Men and Cattel must live upon," and he quotes Genesis 9:23 and 1:29 in this section. Contrast his positive claim with the hedging of a similar claim in which "bare philosophy" has not been supported by revelation, regarding which he concludes, "Whether this be a demonstrative collection I shall not now debate." Nevertheless, bare philosophy is sufficient to show "*That* the Sun, Moon and other Coelestial Bodies, Excellently Declare the Power and Wisdom, and consequently the Glory of God."[35]

Boyle's extensive discussion closes: "*That* all Consideration of *Final Causes* is not to be Banish'd from Natural Philosophy . . . 'tis rather Allowable, and in some Cases Commendable, to Observe and Argue for the Manifest Uses of Things that the Author of Nature Pre-ordain'd those Ends and Uses." He finds that "'tis Warrantable" to consider the parts of animals, "Pre-ordained to such and such Uses, relating to the Welfare of the Animal (or Plant) itself, or the *Species* it belongs to." But he cautions against the capacity of natural philosophy, when not aided by revelation, to elucidate the second among his categories of ends: "from the Supposed Ends of Inanimate Bodies, whether Coelestial or Sublunary, 'tis very Unsafe to Draw Arguments to Prove the Particular Natures of Those Bodies, or the True System of the Universe." This caution about "particular natures" would appear not to apply to lesser claims—the quotation that concludes the paragraph just above is one example—but it does apply to discerning specific purposes from assemblages both large and middle-sized, including worms and wood, where those claims are not supported by scripture.

Boyle finishes with another caution that faintly echoes Bacon's reservation as he writes, "a *Naturalist*, who would Deserve that Name, must not let the Search for Knowledge of *Final Causes*, make him Neglect the Industrious Indagation of *Efficients*." [36] Thus Boyle has chased back Bacon's separation of final causes from physick. The difficulty Boyle saw in divining general providence does not present a bar in principle against such explanation for wood rot. Nevertheless Boyle was skeptical of the degree to which the Naturalist could successfully pursue such explanation. That skepticism concerning topics within his second category would be chased back by others within the Royal Society following his death.

# Physico-theology matures

When Robert Boyle died in 1691 his will included the following provision:

> Fifty Pounds *per Annum* for ever, or at least for a considerable number of years, to be for an annual Sallary for some Learned divine . . . To preach *eight* Sermons in the Year, for proving the *Christian Religion*, against notorious Infidels, *viz. Atheists, Theists, Pagans, Jews,* and *Mahometans,* not descending lower to any *Controversies* that are among Christians themselves.[37]

Boyle had named a board of five trustees for the bequest, including a lawyer, a gentleman who promoted religion in the new world, and two eminent churchmen. The fifth trustee was John Evelyn, a London intellectual and Fellow of the Royal Society who played a crucial role as an agent of Isaac Newton in steering the Boyle lectures. Margaret Jacob and Henry Guerlac have argued that Newton, at the height of his influence, is likely to have suggested either the name of the first lecturer or the specific focus upon physico-theology to Evelyn at a meeting shortly after Boyle's funeral. Even before Boyle's death, Newton would indicate an interest in promoting argument from design that relates to Boyle's second category: David Gregory, a contemporary observer and correspondent of Newton's, indicated that Newton promoted Bentley for the first Boyle Lecture as a counter to John Ray, who focused upon the third category. Guerlac and Jacob write that "Newton was obviously suggesting that his discoveries in celestial physics would serve the argument from design better than that reliance upon the 'contrivances' in animals and plants which John Ray had recently catalogued." [38]

Newton's interest in physico-theology is also intimated at the opening of a frequently quoted letter of 1692: "When I wrote my treatise about our Systeme I had an eye upon such Principles as might work wth considering men for the beliefe of a Deity & nothing can rejoyce me more than to find it usefull for that purpose." Richard Bentley was the recipient of Newton's letter and he would become the first Boyle lecturer in 1693, repeating the role in 1694. Bentley appears to have calibrated his arguments to Newton's instruction, as is apparent from correspondence that includes four letters from Newton to Bentley during 1692-1693.[39] Bentley's final three lectures, collectively entitled "A Confutation of Atheism from the Origin and Frame of the World," particularly fill Newton's bill. The previous three lectures concern "A Confutation of Atheism from the Structure and Origin of Human Bodies," and, as the title indicates, these largely concern the fitness to their uses of the parts of the human body. Bentley is one of several influential physico-theologians who, at the turn of the eighteenth century, set the agenda for its future. One earlier outlier on the continent, François Fénelon's *Traité de l'existence de Dieu et de la réfutation du système de Malebranch sur la nature et sur la Grâce* (1685), contains argument approaching 100 pages that sketched physico-theology. Another author of note is John Ray, whose important empirical work on botany and fossils is complemented by *The Wisdom of God Manifested in the Works of the Creation* (1691).

All of the abovementioned Englishmen were Fellows of the Royal Society and as I have attempted to indicate, most had significant professional and philosophical interconnection, reflecting both association and rivalry. The connections, beginning in the early days of the Society, are bountiful. John Ray's balance of scientific and theological work is reminiscent of that of Wilkins, who put Ray to work on the botanical sections of his *An Essay Towards a Real Character and a Philosophical Language* (1668)—the work noted above as an intellectual precursor to aspects of Locke's *Essay*. Newton promoted Bentley over Ray, and another close associate of Newton's, Samuel Clarke, would follow as another Boyle lecturer a decade later. Clarke would present Newtonian, metaphysical and physico-theological themes in *A demonstration of the being and attributes of God* (1705). The next Boyle lecturer to write a book focused on physico-theology would be William Derham, who achieved a great literary success with *Physico-Theology, or a Demonstration of the Being and Attributes of God from his Works of Creation* (1713). Derham would be among a few eighteenth century physico-theologians whose work experienced popularity, to be followed by the even greater success of Noël Pluche, whose writing could be found in a good portion of well-stocked libraries over the following half-century.[40]

# Conclusion

I have argued that empirically detailed writing in natural theology received a particular boost and a specific modern cast in late seventeenth century England. Francis Bacon separated natural theology from natural philosophy early in the century, but his methodological prescription was quite reversed by authors late in the century, most especially Boyle, who nevertheless retained an affinity to Bacon's objectives by underlining the importance and the priority of inquiry into efficient causes over the search for final ones in science. Boyle was skeptical of success for inquiry into final causes particularly concerning what I have called assemblages, but he saw sufficient additional evidence available for some conclusions within scriptural support. His skepticism, I have argued, was not well supported philosophically, which might serve to explain why it was so thoroughly trammeled by later generations, including Pluche, Biberg and Linnaeus. Indeed, Boyle was open to ignoring the caution himself when he begged license to speculate, even in the passages of the *Disquisition about the Final Causes of Natural Things*:

> I am not averse from thinking, that Humane Ends, (or uses that relate to Men,) may have been designed by God in several Creatures, whose *Humane Uses* Men are not yet aware of . . . And therefore, it cannot sagely be concluded That every thing whose Usefulness to Man is not yet obvious, nay, That every thing that seems hurtful to him, can never be made beneficial to him. . . . *Vipers* are Venomous Animals; but yet their Flesh is a main Ingredient of that famous Antidote *Treacle* . . . As the excessive Rains that cause the over-flowings of Rivers in divers parts of *Africk*, and some other Countries, tho' they seem rather Destructive than profitable, do yet, by their seasonable Inundations, make *Egypt* and some other Countries exceedingly Fertile, that without them would

be very Barren.[41]

Boyle might, then, have quietly held a position concerning providence that Pangloss would have applauded.

The context that produced physico-theology was clearly religious and political. It is unsurprising that a large body of Protestant intellectuals well-placed in a relatively peaceful society with a strong tradition of open speech, would develop links between science and critical discussion of both divinity and the Bible.[42] There were also bounds to the discussion, as Newton, who chose to sit on the sidelines, knew well.[43] Many others on Europe's continent lived much more intimately with religious division as well as the reminder, in 1633, of Galileo's failure to arrange a peaceable arrangement between science and religion.[44] These aspects of the rise of physico-theology have not been the focus of this chapter, which has surveyed the philosophical and social origins found in the English context. Science, philosophy of science and other English philosophical currents—most particularly the theory of ideas and understanding that we are familiar with in its later development by John Locke—were formative for a field that might alternatively have been called "empirical natural theology." Prior shifts in religious sensibility that emptied the Book of Nature of much of its content also prepared the ground. Other philosophical and theological currents not discussed here—most notably theories of divine agency and predestination—and other philosophical trends—the rise of Spinoza's challenge to such natural theology on the continent—also had both shaping and limiting influences upon the field.[45] Finally, philosophers, including natural philosophers, did much more to promote physico-theology than just write about it: Boyle in particular provided a very important launch pad for the further development of an already healthy tradition of natural theology with his named lectureship, which drew the interest of others in the Royal Society, most notably Isaac Newton, and which spawned two of the most influential physico-theological tracts shortly before and shortly after the turn of the eighteenth century.

# Notes

I thank Allegheny College for release time and sabbatical research funds that allowed for the creation of this chapter, written on sabbatic leave in the pleasant environs of the Center for Studies in Religion and Society at the University of Victoria.

1. See, for example, James J. Aimers and Prudence M. Rice, "Astronomy, Ritual, and the Interpretation of Maya 'E-group' Architectural Assemblages," *Ancient Mesoamerica* 17 (2006): 79-96.

2. Noël Pluche, *Spectacle de la Nature* (8 vols., Paris: Veuve Estienne, 1732-1750). Translation is from the 4th English edition, *Nature Display'd*, trans. Humphreys (London: R. Davis *et al.*), Vol. 3: 394. This text and other passages by Pluche that give a taste of his exposition of science may be found in the supplementary readings included in Vol-

taire, *Candide*, ed. Eric Palmer; trans. unknown (Peterborough & New York: Broadview Press, 2009).

3. Voltaire, Chapter 4, 55.

4. See Voltaire, *Candide*, editor's introduction, 24-26. Voltaire gives a quick kick to Gottfried Leibniz as well, in the first sentence of the quotation.

5. Carolus Linnaeus *et al.*, *Select dissertations from the Amoenitates Academicae: a supplement to Mr. Stillingfleet's tracts relating to natural history*, trans. F. J. Brand (London: G. Robinson, 1781), Vol.1: 74.

6. Carolus Linnaeus *et al.*, *Miscellaneous Tracts relating to Natural History, Husbandry, and Physick*, Third Edition, trans. Benjamin Stillingfleet, (London: J. Dodsley, 1775), 98, 123, 123. Stillingfleet attributes all of these texts to Linnaeus as collaborator or sole author and he cites an example in which Linnaeus took as his own words those published under the name of one of his students in the *Amoenitates Academicae*. (v) Some recent scholarship supports that judgment: see John L. Heller, "Notes on the Titulature of Linnaean Dissertations," *Taxon* Vol. 32, No. 2 (May, 1983): 218-252, 245.

7. For an introduction to the Bridgewater Treatises see John Robson, "The Fiat and Finger of God: The Bridgewater Treatises," in *Victorian Faith in Crisis: Essays on Continuity and Change in Nineteenth-Century Religious Belief*, eds. Bernard Lightman and Frank Turner (Stanford: Stanford UP, 1990), 71-125.

8. David Hume, *Dialogues Concerning Natural Religion* (1779). Immanuel Kant, *Kritik der Urteilskraft [Critique of the Power of Judgment]* (1790). Regarding Darwin's contribution, see Adrian Desmond & James Moore, *Darwin* (New York: Norton, 1994).

9. John Wilkins, *Of the Principles and Duties of Natural Religion* (London: Maxwell, 1675), Chapter 4, 39-41.

10. Varro, *Antiquitates rerum humanarum et divinarum libri XLI*, a lost work quoted and summarized in Augustine of Hippo, *City of God*, Book 6. Augustine, *The Works of Augustine*, ed. & trans. Marcus Dods, (Edinburgh: T. & T. Clark, 1871), Vol. 1, 239.

11. Perhaps Varro, and certainly Augustine, would have classified Wilkins' first topic not as philosophical, but as civil theology, and of no use as support for claims concerning God's existence or nature: see *City of God* Book 6, ch. 6. Wilkins' second topic, argument for the existence of God "From the Original of the World," does touch on the eternity of the world, so there might be overlap with Varro's topics in that respect. But it is reasonable to expect that Varro's reference is to argument that concerns the *necessity* of a first cause or prime mover (e.g., Aristotle, *Metaphysics* 12). Wilkins explicitly restricts himself to probable arguments concerning cosmological topics (Ch. 5). His views on whether theology is even capable of necessary argument—of "mathematical certainty," as opposed to "moral certainty"—are difficult to discern, as they are connected to further theological views concerning the possible inconsistency between God's justice in judgment of sinners and a rational compulsion to belief: see Wilkins, Ch. 3, point 5, 30-33.

12. Thomas Aquinas, *Summa Theologica*, Ia Part 2, Article 3. *The Works of St. Thomas Aquinas* Second Edition, trans. Fathers of the English Dominican Province (London: Burns Oates and Washbourne, 1920): Vol. 1.

13. Xenophon, *Memorabilia*, trans. H. G. Dakyns (MacMillan & Co. 1894), Ch. 4.

14. Francis Bacon, *The Works of Sir Francis Bacon* (London: Bayne & Son, 1844), Book II, 96.

15. Bacon, Book I, 46. The italicized quotation is Matthew 22: 29 or Mark 12: 24.

16. Thomas Aquinas, *Summa Contra Gentiles*, IV.1. *The Works of St. Thomas Aquinas* Second Edition, trans. Fathers of the English Dominican Province (London: Burns Oates and Washbourne, 1920): Vol. 15.

17. Supporting biblical passages include Job 12:7-9, Romans 1: 19–20, Psalm 19: 1. Raymond Sebond, *La Theologie Naturelle de Raymond Sebon*, transl. Michel de Montaigne (Rouen: Jean de la Mare, 1641), author's preface, n.p.

18. Sebond, 1, 10-14.

19. William B. Ashworth, "Natural history and the emblematic world view," in *Reappraisals of the Scientific Revolution*, eds. David C. Lindberg and Robert S. Westman (Cambridge: Cambridge University Press, 1990), 303-332.

20. Bacon, *Works*, Book II, 104.

21. Bacon, *Works*, Book II, 106.

22. Bacon, *Works* Book II 46; Book III, 341; Book II, 96, 106, 104; Book I, 8; Book II, 104, 107.

23. René Descartes, *Principles of Philosophy* I. 28. In *The Philosophical Writings of Descartes*, Vol. 1, ed. & trans. John Cottingham, Robert Stoothoff and Dugald Murdoch (Cambridge: Cambridge University Press, 1985), 202.

24. Abraham Cowley in Thomas Sprat, *The history of the Royal-Society of London, for the improving of natural knowledge*, (London: J. Martyn, 1667).

25. Henry Power, *Experimental Philosophy*, ed. Marie Boas Hall (USA: Johnson Reprint Corporation 1966), 190-1.

26. Robert Hooke, *Micrographia, or, Some physiological descriptions of minute bodies made by magnifying glasses with observations and inquiries thereupon* (London: Royal Society, 1665). See pp. 2 (quoted in this chapter), 193-4. Note that Hooke does not refer to his effort as 'physico-theology': the noun appears not to have been used at the time, and the adjectival form 'physico-theological' may not have settled into the specific meaning that attaches it to argument from design before John Ray's *Three Physico-Theological Discourses* of 1693.

27. Wilkins, *Natural Religion*, Chapter 6: 84, 82-4.

28. Wilkins, *Natural Religion*, 102-3, 122-3; quotation 102.

29. Royal Society, "An Abstract of Dr. Wilkins' Essay Towards a Real Character and a Philosophical Language," in *The Mathematical and Philosophical Works of the Right Reverend John Wilkins* (London: C. Whittingham, 1802). John Locke, *An Essay Concerning Human Understanding*, ed. P. H. Nidditch (Oxford: Clarendon, 1975), Epistle, p.10.

30. François Fénelon, *Traité De L'existence Et Des Attributs De Dieu* (1685) in *Oeuvres Philosophiques de Fénelon, Nouvelle Edition*, ed. M. A. Jacques (Paris: Charpentier, 1845), 1; Bernard Nieuwentyt, *The religious philosopher: or, the right use of contemplating the works of the Creator . . . designed for the conviction of atheists and infidels. . .*, trans. John Chamberlayne (London: J. Senex et al., 1718), i.

31. Robert Boyle, *Disquisition about the Final Causes of Natural Things*, in *The Works of Robert Boyle, ed.* Michael Hunter and Edward Davis (London: Pickering & Chatto, 2000), Vol. 11, 81, 83.

32. Boyle, *Disquisition*, 64.

33. Boyle, *Disquisition*, 87; and see a case of Boyle's neglect of middle-sized assemblages at, for example, p. 94. For discussion of Boyle's concerns for teleology vis a vis the third category, the parts of animals (and plants), see James G. Lennox, "Robert Boyle's Defense of Teleological Inference in Experimental Science," *Isis* 74 (1983): 38-52.

34. Boyle, *Disquisition*, 146.

35. Boyle, *Disquisition*, 107, 107-8, 96, 151.

36. Boyle, *Disquisition*, 151. "Indagation of *Efficients*": inquiry into efficient causes.

37. E. Budgell, *Memoirs of the lives and characters of the illustrious family of the Boyles*, Third Edition (London: Oliver Payne, 1737), Appendix, pp. 25-6.

38. Henry Guerlac and M. C. Jacob, "Bentley, Newton, and Providence: The Boyle Lectures Once More," *Journal of the History of Ideas*, Vol. 30, No. 3 (Jul.-Sep., 1969): 307-318; 309, 317.

39. Compare, for example, instruction toward the end of Newton's second letter (January 17, 1693), which appears to have directed Bentley's claims in Lecture 7, p. 36. See pp. 168, 210 of *The works of Richard Bentley*, ed. A. Dyce (Francis McPherson: London, 1836), Vol. 3.

40. Daniel Mornet notes, in *Les Sciences de la Nature en France, au XVIIIe Siècle* (Paris: Armand Colin, 1911), 31-2, that volumes of *Spectacle de la Nature* appeared in over 40 percent of auction catalogs in France between 1750-1780. *Spectacle* had great currency in England, where at least twelve editions in two English translations were published by 1758.

41. Boyle, *Disquisition,* 113, 115, 115, 116.

42. See David R. Oldroyd, *Thinking about the Earth: A History of Ideas in Geology* (London: Athlone, 1996), 47-54.

43. See Stephen Snobelen, "Isaac Newton, Heretic," *British Journal for the History of Science* 32 (1999) 381–419.

44. See Galileo, *Letter to the Grand Duchess Christina*, and note Descartes' hesitation to make available his cosmological treatise, *Le Monde*, in the light of the condemnation of Galileo by Church authorities. Details of this history may be found in Steven Gaukroger, *Descartes' System of Natural Philosophy* (Cambridge: Cambridge University Press, 2002), 19-21.

45. See all of: Peter Anstey, "Boyle on Occasionalism: An Unexamined Source," *Journal of the History of Ideas* Vol.60 #1 (Jan 1999): 57-81. Margaret Cook, "Divine Artifice and Natural Mechanism," *Osiris*, 2nd Series 16 (2001): 133-150. Jonathan Israel, *Radical Enlightenment* (Oxford: Oxford University Press, 2001).

# Chapter 3

# In Defense of "Rationalist" Science

## Anya Plutynski

Mainstream philosophy of science has embraced an "empiricist" approach.[1] To be slightly more precise, I venture that most philosophers of science today would endorse the view that experience is the source of most if not all scientific knowledge. The aim of this chapter will be to challenge the consensus, by showing how we cannot and should not abandon all elements of the "rationalist" tradition, a tradition often identified with philosophers such as Descartes. The very idea that science could be founded on "reason," or in some sense *a priori* seems absurd, at least at first pass. Surely, today, it would be difficult to find anyone who regards intuition or the "natural light" as a source of information about the world. Even logical and mathematical knowledge—once regarded as paradigmatic examples of knowledge founded in "reason"—are now more often than not viewed as a product of experience, or, at very least, a result of combining experience and, perhaps, our psychological dispositions.[2] Further, many argue that explanations in the sciences must be empirical; thought experiments and mathematical truths may not "explain"—or, at least, they do not tell us anything interesting or new about "the world."[3] Appeal to "pure reason" will get us nowhere; there is no longer any plausible role for "first philosophy." According to one popular conception of "naturalism," this is what it means, after all, to be a "naturalist"—and, we are all naturalists, now.[4]

How then might one defend a view so far afield from the consensus? My aim here will be to carve off some of the less palatable rationalist commitments from the elements of "rationalist science" I endorse. Although I am not the first to challenge the traditional characterization of the history of modern philosophy as a stalemate between continental rationalists and empiricists, this will be part of my aim, as well. There are several elements frequently identified with "rationalist" science: questioning of sense experience, and particularly, accepted, or "commonsense" observation, the attempt to rethink the "metaphysical" foundations of one's science (in the broadest possible sense), using either thought experiment, or appealing to demonstrative arguments purporting to establish "necessary" truths (under certain assumptions), often (but not always) using either mathematics or geometry, appeal to the assumption that nature (or some part thereof) is itself "systematically" organized, and the associated assumption that unified theories (in some sense) are preferable, and more generally, appeal to "virtues" not usually considered "strictly empirical," such as simplicity.[5] To some extent, labeling such methods "rationalist" is a bit artificial; both "empiricist" and "rationalist" thinkers questioned sense experience, used thought experiment, and appealed to simplicity. Indeed, this is (in part) my point; by using such methods, one does not depart from an ideal. Such "rationalist" tools are good ones for moving science forward; Descartes' ladder (or, perhaps, van Fraassen's "inferential wand") should not be thrown to one side. More detail about why will become clear as we progress; it's best explained by looking to examples.

The structure of the argument will be as follows. First, I will begin by discussing how Descartes exemplified some of the above virtues. Descartes stood at center stage in the history of science, and not simply in the history of philosophy.[6] His scientific work transformed physics; even if he was wrong about many of the conclusions he arrived at, he served as an interlocutor in absentia for much of subsequent physics. Second, I will discuss how and why a rationalist vision for science is not so counterintuitive as one might think; I illustrate several examples of scientists deploying "rationalist" methods to advance science.

# Part I: Rationalist Virtues

## Skepticism about the senses

Of course, what it means to be a "skeptic" about sense experience has varied over time, and has served different purposes in the history of philosophy and of science. Doubt of sense experience is a long-standing tool used by philosophers (or, if you like, "natural philosophers"—i.e., scientists) going back at least to the Pre-Socratics, for raising problems of knowledge. Skepticism, whether mild or radical, is a way of illustrating the underdetermination of belief by sensory experience. Whether the world is, in fact, exactly as it appears, was a question raised by "empiricists" and rationalists alike, from Aristotle to Hume. Galileo uses many vivid examples to demonstrate how our "common sense" perception can fail us. For example, Galileo addresses "common sense" Aristotelian objections to the idea that the earth moves by pointing to a series of theoretical scenarios where compound motion would be perceived as a single directional motion for a person who shared the motion of the object. (Tossing a ball in the hull of a moving ship, or on the back of a galloping horse, one would only observe the ball rise and fall, not the arc of forward motion combined with vertical motion.) Descartes, just like Galileo, argues that relying upon what one is "taught by nature" to discover natural kinds or true explanations is dubious. While he argues that "there is no doubt that everything I am taught by nature contains some truth," truth must wait upon the "intellect" to "examine the matter."[7]

One of Descartes' most vivid examples, from the *Meditations*, concerns the size of the sun relative to the earth; relying upon direct visual inspection to determine the sun's actual size may be misleading. We must (ultimately) resort to geometrical reasoning, to conclude that, in fact, the sun is many times larger than the earth. For our purposes, it will be best to show how Descartes viewed skepticism, or perhaps better, doubt, as a tool. Doubting the senses was, for Descartes, the first step along the way toward a discovery of foundational truths. Descartes believed that a new physics required a new metaphysics. Descartes' skepticism about the senses is thus not merely a first step along the way towards discovery of the "clear and distinct" metaphysical truths he takes to ground his new science; it is also a "rearguard" action, if you like, against his opponents. His goal is not simply to motivate questions about the nature and possibility of knowledge, but also, to question a worldview. In order to do this, Descartes needs to question what are regarded as "common sense" observations about the kinds of things there are in the world (men, trees, stars, etc.), and how we are to explain their behavior (form, matter, "natures").

At the end of the *Meditations*, Descartes notes that our ideas of external things are produced without our consent and cooperation (a cancer diagnosis, or a burned hand). Since we know (by *Meditation* 3, at least), that God is not a deceiver, our perceptions of these external things must be (when properly disciplined by reason) about the world, for

> [God] has given me a great propensity to believe that [ideas of corporeal things] are produced by corporeal things. So I do not see how God could be understood to be anything but a deceiver if the ideas were transmitted from a source other than corporeal things. It follows that corporeal things exist. They may not exist in a way that exactly corresponds with my sensory grasp of them, for in many cases the grasp of the senses is very obscure and confused. But at least they possess all the properties which I clearly and distinctly understand, that is, all those which, viewed in general terms, are comprised of the subject matter of pure mathematics.[8]

Descartes makes a number of points here—our sense experience is "about" the world, but our senses' grasp of the world is, in many cases, "obscure and confused." Here Descartes appeals to some important distinctions: "obscure and confused" versus "clear and distinct" ideas, things grasped via "sensory" experience versus "viewed in general terms." These distinctions are supposed to provide readers with clues as to which kinds of knowledge are trustworthy, at least at this stage in his analysis. His point, put simply, is that the senses may often deceive us. In contrast, the "subject matter of pure mathematics"—a field that includes truths of geometry, for Descartes, as well—is known with certainty. What is the relationship between these two? The clue comes with Descartes use of the expression, "viewed in general terms"; it is only when we abstract away from the appearances that we gain general, and more importantly, accurate understanding of the "real" properties of objects—i.e., properties of objects that truly belong to the object, and are not simply a byproduct of our senses—such properties will include size, shape, and motion. A few paragraphs after this, Descartes makes an important distinction between things "taught by nature" and things "taught by reason." The former are aspects of corporeal things that are particular—such as sensations of sound, light, pain, or flavor. "Habitual" judgments of the sense "inform the mind of what is beneficial or harmful for the composite of which the mind is a part."[9] The trick is not to assume that the perceptual categories that we take for granted (sweetness, brightness, etc.) are what "ultimately" explains the phenomena. This is the fault that Descartes (fairly or not[10]) attributes to the "School metaphysicians," that there is "something in the fire which resembles heat":

> If you find it strange that I make no use of the qualities one calls heat, cold, moistness, and dryness, as the philosophers [of the schools] do, I tell you that these qualities appear to me to be in need of explanation, and if I am not mistaken, not only these four qualities, but also all the others, and even all of the forms of inanimate bodies can be explained without having to assume anything else for this in their matter but motion, size, shape, and the arrangement of their parts.[11]

Descartes' argument is that perceived properties of bodies may "correspond" to "real" properties but not "resemble" them—for instance, sweetness may "correspond" with geometrical features of sweet objects'—e.g., the size, shape, and motion of their parts may correspond to their ability to cause sweet sensations.[12] In other words, Descartes' project in questioning the senses is not, ultimately, to show that *all* sense experience is fundamentally flawed. It is, at least in part, to challenge (what he took to be) Scholastic conceptual categories. That is, his argument is that qualities of heat, cold, moistness, are not "natural kinds." His claim is thus not that our sense impressions of color, flavor, etc., are altogether useless, but that our psychological judgments (what we are "taught by Nature") about the sweetness of a piece of fruit are exactly spot on, at least, for the purposes of our survival. They are, as it were, coded instructions by God to continue to enjoy this tasty mango. The mango is not a delusion, or caused by an evil demon. It is real, and it tastes good because God wants us to eat it. Of course, ultimately, he will argue that the true nature of the mango is that it is an extended, changeable thing, made of parts with definite sizes, shapes, and motions; knowledge of these mechanistic physical properties of the mango will be necessary for the future physics of mangoes (as well as of the future of flavor technology). However, we need not know this in order to know that it's sweet and juicy, and just the thing for a nice afternoon at the beach.

The function of the senses, in short, is to convey information for preservation of the body. What is flawed is not sense experience per se, but habitual judgments about the senses—e.g., that the world is exactly as it appears to us, uncorrected by reason, and prompted, in part, by a good dose of skeptical reflection. The idea that secondary properties (the sweetness or orange color of the mango) are really "in" the thing itself, is the view that Descartes aims to displace. In other words, Descartes' aim is not to reject experience altogether, but to show that the way we ought to do science is to replace our "common sense" observations with a more "objective" understanding, i.e., to use reason to correct "habitual" empirical judgments.

Thus, the senses provide us with knowledge of particular ways in which we can and should come to approach and avoid pain and pleasure, but the "intellect" discerns general metaphysical properties of those particulars—their substance and modes—e.g., size, shape, and motion. But how does the intellect discern these particulars?

## Reason as a Source of Information about the World

Descartes viewed reason as a source of knowledge. This much is uncontroversial; what is far more controversial is what it means to say that "reason is a source of knowledge." A narrow sense of this would have it that knowledge founded on "reason" simply means all and only "a priori" propositions and claims demonstratively following from these. Is this what Descartes intended by use of "reason" as a way of knowing about the world?

No. One traditional reading of Descartes had it that he hoped that all of science should exactly follow the example of Euclid's *Elements*—a deductive demonstration of all of science from first principles that were known a priori. Anything less than perfect certainty was unsatisfactory; "scientia" is all and only "certain and evident knowledge." While Descartes may, early in his career, have hoped that the sciences might follow Euclid's *Elements in achieving the same kind of certainty*, it's clear that he did not believe—even then—that all of science could be deduced (in the narrow sense associated with the "syllogistic art") from a priori first principles.[13]

We must be very careful about the use of the term "demonstration." Descartes made a distinction between "proof" and "explanation"—or, what he called "a priori" justification, versus "a posteriori" justification. The former moves from causes to effects, and is somewhat closer to our notion of demonstration as deduction (but not the same); the latter moves from effects to causes. The word, "demonstrate," Descartes insists, can be used both for proof and explanation, "according to common usage and not in the technical philosophical sense."[14] So, when Descartes claims that he has used his reason to "demonstrate" certain truths, he is not (at least not for the most part) saying that he has proven truths by syllogistic reasoning from a priori first principles. Even in the a priori case, Descartes appeals to a somewhat broader sense of "demonstration" than what is called to mind by twentieth or twenty-first century philosophers.

Descartes was aware that deductive demonstrations (in the sense in which syllogistic arguments are "deductive") prove no more than what is contained in the premises. Thus, early on in his career, he is at pains to distinguish his sense of "deduction" from that of the "logicians": "The syllogistic art is of practically no assistance in the search for truth . . . logicians can form no syllogism which reaches a true conclusion unless the heart of the matter is given, unless they previously recognized the very truth which is thus deduced."[15] Having been schooled by Jesuits, Descartes was certainly aware of this classical objection to the possibility of knowledge via demonstration, discussed in the *Posterior Analytics*.[16] Purely "syllogistic" reasoning as a form of knowledge does not bequeath anything *new*. As Descartes hopes to reform the sciences, indeed, to offer a new, systematic account of the world as a whole, his (intended) sense of "demonstration" is not merely "syllogistic" reasoning.

As Descartes developed his scientific program, his early ideals about the nature and possibility of certainty in the sciences were jettisoned at least in practice, for what we would today regard as a much more probabilistic mode of argumentation.[17] In his empirical work, he makes appeal to the latter sense of "demonstration"—i.e., a posteriori demonstrations from effects to causes. Along the way, he appeals to parsimony, coherence, and unifying power of his theoretical framework; in short, a sense of "demonstration" that falls very far short of Euclidian proofs. Descartes argued with critics that this did not constitute a "failure"; he wrote to his colleague Mersenne:

> You ask if I regard what I have written about refraction as a demonstration. I think that it is, in so far as one can be given in this field without a previous demonstration of the principles of physics by metaphysics, and so far as it has ever been possible to demonstrate the solution to any problem of mechanics, or optics, or astronomy, or anything else which is not pure geometry or arithmetic. But to require me to give geometrical demonstration on a topic that depends on physics is to ask the impossible. And if you will not call anything demonstrations except geometers' proofs, then you must say that Archimedes never demonstrated anything in mechanics, or Vitellio in optics, or Ptolemy in astronomy. But of course nobody says this. In such matters people are satisfied if the author's assumptions are not obviously contrary to experience and if their discussion is coherent and free from logical error, even though their assumptions may not be strictly true.[18]

This passage provides a useful way of explicating how Descartes (at least at this point in his career) viewed "geometrical" demonstrations as distinct from those to be found in the sciences. The former inherit their certainty from the certainty of their first principles. The latter arguments, Descartes seems to suggest, are "demonstrative," to the extent that they rely upon assumptions not "obviously contrary to experience," and, when they are "coherent," and free from "logical error," they are even better.

While it may sound odd to our twenty-first century ears to suggest that demonstration could be a matter of degree, for Descartes, who is attempting to found an altogether new science, appeal to such criteria as "coherence" was an important first step to jettisoning Aristotle's metaphysics and physics. He believed, moreover, that a new physics required a new metaphysics; since, for him, a metaphysical foundation was prior to, and foundational for, the new physics. Descartes wished for his new theory to be persuasive—what does persuasion require? Descartes explains in a letter to Vatier:

> [I]ndeed it is not always necessary to have *a priori* reasons to convince people of a truth. Thales, or who ever it was who first said that the moon receives its light from the sun, probably gave no proof of it except that the different phases of its light can be easily explained on that assumption. That was enough to ensure that from that time to this his view has been generally accepted without demur. My thoughts are so interconnected that I dare to hope that people will

find my principles, once they have become familiar by frequent study and are considered all together, are as well proved by the consequences I derive from them as the borrowed nature of the moon's light is proved by its waxing and waning.[19]

Descartes' appeal to the example of the moon is illuminating, and not only because the pun is so tempting. Descartes is appealing to two " rationalist" virtues here; first, he speaks of his thoughts being "interconnected" and of his principles being "considered all together"; second, he speaks of these very principles being "proved by the consequences I derive from them." In other words, he seems to suggest here that "proving" his principles is not a matter of simple deduction of the phenomena from them, more geometrico, but (at least in part) showing how they are all systematically connected and that they have enormous explanatory power. Or, proof (in a far broader sense than the syllogistic one) may run in the opposite direction of a Euclidian one—the effects may prove the causes. The borrowed nature of the moon's light is "proven" by its characteristic patterns of waxing and waning.

In short, Descartes has stumbled upon the problem of underdetermination. "Descartes' ladder" is this: he scaled the heights of generality, arriving at the simplest possible first principles of physics. Then, he descended to the physical world, and found that he often had to resort to hypothetical explanations. When challenged by his contemporaries, he found that he needed to bridge the gap between the explanations he wished to offer at the level of the "visible world" and the first principles of his physics. How did he know that this particular explanation was preferable to others? How might we bridge such a gap? Descartes offered a variety of criteria—coherence, consistency, simplicity, and explanatory power. As in the case of Thales, Descartes asks his readers to consider what "further observations" one's hypothesis entails—i.e., imagine other plausible consequences of one's hypothesis, and check whether these consequences are borne out by experience. If they "explain" systematically, so much the better.

Further, he claims that no alternative hypothesis could explain the same phenomena as parsimoniously:

Finally you say that nothing is easier than to fit a cause to an effect. It is true that there are many effects to which it is easy to fit many separate causes; but it is not always so easy to fit a single cause to so many different effects, unless it is the true cause which produces them. There are often cases in which in order to prove what is the true cause of a number of effects, it is sufficient to give a single one from which they can all clearly be deduced. I claim that all the causes of which I spoke belong to this class . . . Compare my assumptions with the assumptions of others . . . Compare all their real qualities, their substantial forms, their elements and countless other such things with my single assumption that all bodies are composed of parts. This is something which is visible to the naked eye in many cases and can be proved by countless reasons in others. All that I add to this is that the parts of certain kinds of body are one shape rather than another . . . Compare the deductions I have made from my assumption—about vision, salt, winds, clouds, snow, thunder, the rainbow, and so on—with what others have derived from their assumptions on the same topics.

I hope this will be enough to convince anyone unbiased that the effects which I explain have no other causes than the ones from which I have deduced them. None the less, I intend to give a demonstration of it in another place.[20]

Though one may whimsically assign causes to effects, if there is one cause that explains many effects, Descartes claims, it is more likely the true cause. In this exchange, Descartes invites Morin to compare his assumptions with those of others, proceeding to list the scholastic notions of real qualities and substantial forms. He indicates that his assumptions are more parsimonious—in effect, only that all bodies are composed of parts. Finally, he gives a promise (unfulfilled) that he will be able to give a "demonstration" elsewhere, in his *Principles of Philosophy*.

Yet, here too, there is a gap between the project of the first two parts of the *Principles*, which cover the most basic metaphysical and epistemological questions, and the latter, incomplete parts, which are concerned with particular effects—the visual world. These latter parts often involved "hypothetical" demonstrations. At the end of the *Principles*, Descartes defended the use of idealized and hypothetical suppositions. Descartes remarks, again, in the *Principles*, "In fact, it makes very little difference what initial suppositions are made."[21] And, "We are free to make any assumption on these matters with the sole proviso that all the consequences of our assumptions must agree with experience."[22] The problem, which Descartes acknowledged, is that many general theories were possible; "experience" can be made to agree with any number of suppositions. There was, in short, a gap between his first principles and the concrete applications. Here, Descartes attempts to offer a reply to this problem:

[I]f people look at all the many properties relating to magnetism, fire, and the fabric of the entire world, which I have deduced in this book from just a few principles, then, even if they think that my assumption of these principles was arbitrary and groundless, they will still perhaps acknowledge that it would hardly have been possible for so many items to fit into a coherent pattern if the original principles had been false.[23]

Descartes confesses that the best argument available to him is an inference to the best explanation. It is implausible, he claims, that his assumptions were groundless. For, consider all the phenomena that he was able to explain on these assumptions! His explanation was comprehensive, systematic, and unifying. Not only is his view, in contrast to Aristotle's and the Aristotelians, simpler, it is, he claims: a better explanation of the phenomena. Why? He makes a few simple assumptions, such assumptions are consistent with observation. Moreover, these assumptions are proved by "countless reasons," not least of all, they fit the observations; they fit a "coherent pattern." His theory has the virtues of parsimony, unifying power, AND empirical adequacy.

# Part II: Principles of Rationalist Science

The aim of this chapter has been twofold. First, I've reviewed (if briefly) some of Descartes' "rationalist" methods. Second, I wish to show how Descartes anticipates modern science in important ways. I'd like to expand on this second goal further here, by giving examples of what I would call a "rationalist" dimension to the sciences—one all too frequently overlooked. How is a devout rationalist in any way like a "modern" scientist?

Modern scientists do not consult their intuition to discover a priori truths about the natural world. Moreover, unlike Descartes, who seemed to have deep reservoirs of faith in his abilities to get things exactly right, most modern scientists are fallibilists.[24] Nonetheless, I believe a good case can be made that modern scientists do deploy "Descartes' ladder" to good effect in attempting to bridge the gap between what are often highly general and idealized models and claims about particular effects in nature. It's well known that a good part of modern science consists in the development of mathematical models, and the generation of arguments founded on these models. In many of these cases, arguments have the form of, "If we assume X, Y, and Z," where these are idealized assumptions about the world, "then, this will follow." These arguments involve treating the world as an idealized system with properties, which we (often) know not to hold (at least not exactly). And, many of the arguments in these cases are demonstrations of what "must" be the case (at least if you accept such assumptions). Such arguments are, I would argue, often necessary for new theories in the sciences to even begin to lift off the ground. They serve, both, as "tests" of our "commonsense" assumptions, suggesting alternatives that we may not have considered, perhaps because we are entrenched in a theoretical framework. Moreover, such model building and associated thought experiment often functions as a way of rethinking the "metaphysical foundations" of our science. Such models *frame the space of possibilities*; without knowing what is "possible," even in admittedly "idealized" systems that do not directly describe the world as it is, we cannot begin to frame some of the questions we might seek to test. The use of "what if" here is not "merely" precursor or heuristic; it is, I would argue, part of the process of scientific argumentation, and scientific innovation—if you like, "first philosophy" in the sense of challenging our most basic assumptions. Some of the most important theoretical innovations in the biological and physical sciences occurred through the process of generating such idealized models or thought experiments.

It's what happens next that most philosophers of science seem to focus their exclusive attention upon; novel empirical discoveries (which can, by the way, occur only once we know what we "should" be looking for, once we have the theoretical framework to make sense of them) confirm our theories. Yet, this process of confirmation is rarely so spontaneous as the newspaper headlines or textbooks would have us believe. As most "post-Kuhnian" philosophers of science know, choice among competing hypotheses is often (if not always) a process of negotiation. The new theory or hypothesis needs to be shown to be superior.

Here again, I would suggest, methods of argumentation analogous to those of Descartes play a significant role in modern science. Rationalist science, I would argue, involves several features. First, doubt of what are, at the time, taken to be "common sense" observations—accepted, empirically informed views of the way the world is—can be a useful first step toward a new theoretical paradigm. Second, offering up a "theoretical demonstration" or thought experiment—i.e., what "must" be true—though, only under certain assumptions. Third, and finally, a demonstration that such a theory better explains a wider range of phenomena, is simpler, and/or is more coherent with other well-confirmed theories or empirical observations. An example is Mendel—at the time Mendel did his famous experiments, most assumed that heredity was blending—i.e., that the hereditary material was not "discrete," but was, instead, rather like the combining of ingredients in a soup, or dye in a vat. The offspring of a tall parent and a short parent would thus be of average height. This was simply "common sense"; hundreds of observations confirmed it. Indeed, most hereditary traits are quantitative—height and weight vary not in kind, but by degree, and offspring generally do tend to regress to the mean of their parents. So, Mendel, in idealizing heredity as a matter of discrete particulates, was countering common sense observation. This was necessary to generate his laws of segregation and independent assortment. Though we now know that heredity is far more complex than Mendel's simple picture suggested, Mendel's step was essential to subsequent developments of both genetics, and evolutionary theory. His use of examples of "rare" cases to prove the rule was an ingenious move; his use of qualitative traits—traits that vary discretely—made possible a quantitative theory of heredity. Both his challenge to "common sense" observation, and his deployment of an entirely new metaphysics of heredity are common "rationalist" moves.

"Demonstration" in the broader sense of showing what one might explain both new and old phenomena, or unify what are viewed as competing paradigms, under a new theoretical framework is a common rationalist strategy. Consider, for instance, Fisher's demonstration of the consistency of Biometric patterns of inheritance with Mendelian theory of heredity. In order to show that these two theories were consistent, he had to idealize away from features of each theory—he assumed, for instance, that there is no dominance, and that single phenotypic traits are caused by many hereditary particles of equal effect. In Fisher's case, as in Mendel's, what was important was the demonstration of what is "possible," even on assumptions known (or believed) to be false. The (idealized) Mendelian assumption turned out to be explanatory (in Descartes' sense of being capable of deriving the phenomena from it—the light of the observations reflected back on the theory), and this explanatory power made possible the adoption of a theoretical framework, which (however limited) unified and advanced genetics and evolutionary biology for the next fifty years. Whatever we now think of the limitations of the "Fisherian" model of inheritance and evolution, the theoretical framework needed a "foothold" —i.e., what was important in Fisher's demonstration was that he proved that, at least on some assumptions, Mendelism and Darwinism were (at least) consistent with one another.

Some have argued that appeal to "rationalist" considerations—e.g., unifying power, and simplicity—runs counter to good science. Since false theories may be unifying, and simple theories may be false, some argue that such virtues are either at variance with good science (van Fraassen), or, alternatively, reduce to or are supported by empiricist warrant—e.g., predictive accuracy. For instance, in a recent paper, Elliot Sober explains how unifying power may be relevant to predictive accuracy, in some contexts.[25] We ought to prefer models that are more unifying (or, more precisely, models with fewer adjustable parameters), Sober argues, because they are more likely to be predictively accurate. Sober deploys Akaike's theorem as a useful way of solving what he calls "Peirce's problem." Others contend that "rationalist" virtues like unification might be recaptured in Bayesian terms.[26] The problem is a very old one; how do we move from the observations to an abstract theory? Peirce was concerned with the rules that govern inference to the best explanation. Peirce's question was, in a way, not unlike Descartes' question viz. Scholastic metaphysics: Which of two competing theories would lead to a total system of beliefs with the most explanatory power? Why ought we to choose, of two competing explanations, the one that better unifies the phenomena?

For Sober, the Peircian problem is just the problem of model-selection in a different guise. The problem of model-selection is familiar from the case of curve fitting. Scientists prefer a model that is simpler, or one that is described by an equation with fewer parameters rather than more. Why? According to Sober and Forster, the model with fewer parameters is more likely to be predictively accurate. Sober and Forster deployed Akaike's theorem in the context of the curve-fitting problem, and they argue that the same theorem explains why we ought to prefer a theory that is more unified.[27]

Sober asks us to imagine that we seek to explain the effects of smoking on rates of cancer. He describes two models, one "unified" and one "disunified." The first model, "U," simply ascribes the rates of cancer to tobacco use, regardless of the brand of cigarette. The second model, "D," treats the effects of smoking two different brands of cigarettes as distinct. Sober argues that it's inevitable that the second model will fit the data better, but contends that "U" is nonetheless preferable. According to Akaike's theorem, an unbiased estimate of a model M's predictive accuracy is approximately equal to the log-likelihood [L(M)]-K. Or, the log of the likelihood of the fitted model (a model whose parameters are estimated by maximum likelihood from some data set), minus the number of adjustable parameters, is a good estimate of the predictive accuracy of the model. In other words, a model with fewer adjustable parameters—the "unified" model on Sober's reading—will be more predictively accurate, and so preferable, in the longer run. So, according to Sober, "If you want to find out which of several models will be predictively accurate, then it is an objective fact that unification is relevant."[28]

Likewise, Myrvold and McGrew both give a Bayesian re-formulation of unification as a virtue; essentially, their argument is that appeals to "unifying power" can be reduced to empiricist virtues. By deploying Bayes' theorem, they show that the greater positive relevance of facts to one another, or, the higher probability of all the facts combined on a unifying as opposed to the disunified hypothesis, ceteris paribus, lends greater posterior probability to the unifying hypothesis.

Sober essentially assimilates the problem of choosing between more or less unified theories to the problem of model selection. Likewise, Myrvold and McGrew reduce the value of unification to Bayes' rule and empiricist virtues. Both of these strategies are certainly brilliant solutions to the problem of valuing unification for the empiricist. However, I wish to suggest that the search for empiricist justification of unifying power is rather post hoc, at least from the perspective of the history of science. When we have two theories, and the evidence does not (at least not yet) allow us to choose, what ought we to do? Does "unifying power" give us warrant to prefer one to the other? Sober, McGrew, and Myrvold answer yes, but only BECAUSE the unified theory will be (under certain conditions) more empirically accurate, ceteris paribus. But, unifying power and predictive accuracy surely do occasionally pull apart, and it's not always clear that when they do, we should prefer the latter to the former.[29] Sometimes, I would suggest, having a general explanatory theory, especially in early stages of a science, is preferable to one that is predictively accurate. That is, having a theory that either unifies theoretical frameworks otherwise viewed as inconsistent, or a theory that unifies diverse phenomena under a new metaphysical framework moves science forward in important, novel ways. While I cannot claim this as a general rule, it seems that there are enough cases of this in the history of science to bear consideration.

For instance, classical population genetic theory unified a diversity of phenomena under a common set of mathematical models. But, classical population genetics is infamously bad at prediction. The kinds of generalizations that it generates are useful. They describe what's possible and what's necessary in evolving populations, given certain assumptions. They are, in short, rather like a very abstract calculation device, for imagining what evolution would be like in populations where genes determine traits, among other (idealized) assumptions. One may derive, for instance, broad claims about the relative significance of drift v. selection, say, in populations with x population size and y selection coefficients. But, such predictions and models are highly idealized and general; it's quite difficult to "apply" them—yet, the theory served a central explanatory role in the history of evolutionary biology. They provided a new way of "seeing" populations—as changing over time due to a set of "forces" much like Newton's physics. The role of population genetics is to provide a dynamic way of representing populations over time, at a time when very little was understood about the relationship between genes and traits, much less what genes were. Given all the complexity and difficulty of representing the many features potentially relevant to evolutionary change—heredity, ecology, chance, phenotypic adaptation, etc.—a simplified model provided a way to unify complex, diverse populations under a single explanatory framework. Even if today, the "extended" evolutionary synthesis will require a broader set of tools to explain the evolution and diversity of life, this framework was, I would suggest, a simple, but unifying and necessary first step toward seeing evolution as a process that could be explained using quantitative models and general causes (causes that operate in all populations—selection, mutation, etc.).

Arguably, Descartes made a similar move in his physics of size, shape and motion. Though his physics failed, it failed spectacularly; seeing all of the physical world as explainable in terms of matter in motion made way for Newton's physics.

Finally, contra Sober, it's not clear that every case of what we consider a more unified theory involves choice of a model with fewer rather than greater parameters. For instance, at the turn of the century, biologists were seeking single factor explanations of evolution (e.g., neo-Lamarckism, mutationism, etc.). What the early synthesis of Biometry and Mendelism and the development of theoretical population genetics made possible, however, was the realization that no single factor need predominate in evolution. Biometers believed that evolution was governed primarily by gradual selection on minor variations; Mendelians believed that evolution was a product of major mutations. The models of theoretical population genetics take no single factor to be primary in generating evolutionary change. Rather, the models integrate the different factors of selection, mutation, migration, and drift in a common mathematical framework. Population genetics is thus in a sense unifying, but not because it requires fewer parameters than the alternatives then available. Rather than take one factor (mutation, selection) to be the exclusive cause of evolution, several were taken to work in concert (though some were considered of greater or lesser significance). The virtue that could be claimed by theoretical population genetics is that it "unified" a variety of processes under a single theoretical framework, however idealized, and was more general as a quantitative description of evolution in populations than any "single factor" alternative.

This dialectical process of reasoning—moving between one's own model, the going alternatives, and the world—and showing that one's own preferred hypothesis explains "more," and "more simply"—is often what takes place exactly during those times of "negotiation" when it is not clear what the "evidence shows." Such negotiation is, I would argue, not "bad science"; all too often, the data is consistent with a variety of hypotheses, and thus the goal is to show how a framework both is consistent with observations, but also (a) coherent, or logically consistent, (b) adopts simpler assumptions—i.e., assumptions that while not strictly true, are at least, in Descartes' words, not "obviously contrary to experience," (c) is more comprehensive. The process of exploring this, and the "rationalist" arguments appealed to in this context, often help "unpack" as it were, the assumptions of alternative theoretical frameworks, and show how they may be displaced, if not once and for all, at least in theory. This unpacking is part of the process of the advance of science.

Recall the three elements of Descartes' view outlined above. First, he argues that direct observation should be questioned and corrected by rational reflection. Moreover, he wished to replace Scholastic metaphysics and provide a new metaphysical foundation for physics. Second, Descartes offered a simple, "streamlined" alternative to his competitor; and, he often asked his readers to imagine scenarios which were false. In the end of the *Principles*, Descartes requests of his readers to consider the possibility that his results, "will be allowed into the class of absolute certainties,"[30] though he makes a much better case for a weaker standard, "moral certainty" on the basis of arguments from simplicity and explanatory power. For all he can say is that his mechanistic assumptions seem to explain a great deal.

Descartes' *Meditations* is frequently taught as the statement of a theory of knowledge fundamentally at odds with Hume's *Enquiry* or Locke's *Essay*. Alas, Descartes is often made out to be the (rather naïve) loser in this conversation. However, Descartes was far from naïve. It is true that Descartes believed that there was a rational faculty by which we could intuit clear and distinct ideas about the nature of God, the truths of mathematics, and the truth of a substantial distinction between mind and body. The Cartesian circle is, indeed, circular. My aim is not to challenge this. Rather, I seek to identify what, of Descartes' actual scientific argumentation is retained in much of science today, and might reasonably be called "rationalist" science. Descartes did anticipate modern science, but in ways that are not usually acknowledged by us moderns; he was modern in several ways, not only in rejecting Scholastic physics and metaphysics, and not only in his "mechanistic" physics, but also in advancing a set of tools for rationalist science.

What is rationalist science? It is not (or, not exclusively) to be committed to a set of foundational, metaphysical truths, said to be known via reason alone with absolute certainty. Descartes went very far beyond simply asserting that bodies are made up of parts with size, shape, and motion. What is much more interesting is how he deployed these assumptions in offering explanations of motion of bodies. Rationalist science, as Descartes actually practiced it (instead of what he avowed) involves questioning appearances, rethinking metaphysical foundations, using either thought experiment or idealized assumptions, deriving "necessary" conclusions under such idealized assumptions, and comparing one's model with alternatives, to see whether it can explain more, with fewer (suspect) assumptions. Of course, today, such a set of strategies will be familiar to empiricists and rationalists alike. My suggestion here has been that attempting to reduce such a process to seeking empirical adequacy loses a great deal of the texture of the history of science, in a way that philosophers should be wary of. Reducing such strategies to empiricist virtues, while perhaps a promising solution to the "problem" of the value of unification, is post hoc. While we may, in some cases, reduce or explain away the value of "unified" theories, unification is a many splendored thing.[31] "Rationalist" methods such as those discussed above have played an important role in the history of science, to the extent that it's difficult, today, to say which count as "rationalist" and what as "empiricist." Rationalists like Descartes were often wrong, but fruitfully so; there is value to "systematic" and "foundational" rethinking in science, so perhaps naturalists need not give up "first philosophy" in the broadest possible sense.

# Notes

1. This sense of empiricism is rather different from use of the term "empiricist" understood in contrast with "scientific realism," which concerns the ends of science, rather than the methods. While I have not conducted a survey, a search of "PhilPapers Survey" for "Target Faculty" in General Philosophy of Science shows that over half (56.3 percent) would categorize themselves as "empiricist." See http://philpapers.org/surveys/ (accessed July 10, 2010).

2. For a much more detailed discussion of the source(s) of mathematical and logical knowledge, see Maddy, P., *Second Philosophy: A Naturalistic Method.* (Oxford: Oxford University Press, 2007).

3. See, e.g., Norton and Brown for an exchange, in Hitchcock, C. *Contemporary Debates in Philosophy of Science* (Oxford: Blackwell, 2004).

4. Or, at least 49 percent of all respondents endorsed naturalism, but 64 percent of all specialists in general philosophy of science. See http://www.philpapers/surveys/ (accessed July 10, 2010).

5. Some, but not all of these are mentioned by Stump, D., "Rationalist Science" in *Blackwell's Companion to Rationalism* (Oxford: Blackwell Publishing, Ltd., 2005).

6. See Shea, W. R. *The Magic of Numbers and Motion: The Scientific Career of René Descartes* (Canton: Science History Publications, 1991); Garber, D. *Descartes' Metaphysical Physics* (Chicago: University of Chicago Press, 1992); Ariew, R. *Descartes and the Last Scholastics* (Ithaca: Cornell University Press, 1999); Gaukroger, S., J. Schuster, and J. Sutton, eds. *Descartes' Natural Philosophy* (London: Routledge, 2000) Gaukroger, S. *Descartes' System of Natural Philosophy* (Cambridge: Cambridge University Press, 2002); Hatfield, G. "Science, Certainty and Descartes." Pp. 249-262. in A. Fine and J. Leplin, eds. *Philosophy of Science Association* vol. 2 (East Lansing: Philosophy of Science Association, 1988); Hatfield, G. *Routledge Guide to Descartes' Meditations* (London: Routledge, 2002); Slowik, E. *Cartesian Spacetime* (Dordrecht: Kluwer, 2002).

7. Cottingham, Stoothoff and Murdoch trans. *Philosophical Writings of Descartes* (3 vol) (Cambridge: Cambridge University Press, 1995). II 82-83.

8. Cottingham, *Descartes*, II 80.

9. Cottingham, *Descartes.* II 83.

10. Thanks to E. Schleisser for making this point about Descartes rather uncharitable reading of his predecessors. See also, Grant, *God and Reason in the Middle Ages* (2001), and Grant and Murdoch, *Mathematics and its Applications to Science and Natural Philosophy in the Middle Ages* (1987).

11. Cottingham, *Descartes* I 25-6.

12. Cottingham, *Descartes* II 81.

13. For a discussion, see Garber, 1978.

14. Cottingham, *Descartes* III 198.

15. Cottingham, *Descartes* I 406.

16. Book 1, Chapter 3, 72b.

17. For further discussion, see Garber, 1992; Hatfield, 1988; Raftpoulos, 1995.

18. Cottingham, *Descartes* III 141.

19. Cottingham, *Descartes* III 564.

20. Cottingham, *Descartes* III 199-200.

21. Cottingham, *Descartes* I 103.

22. Cottingham, *Descartes* I 101.

23. Cottingham, *Descartes* I 205.

24. Thank you to David Stump for pointing this out.

25. Sober, E., "Two Uses of Unification." Pp. 205-216 in *The Vienna Circle and Logical Empiricism: Re-evaluation and Future Perspectives*, edited by Stadler, F. (Kluwer: The Netherlands, 2003).

26. See McGrew, T. "Confirmation, Heuristics, and Explanatory Reasoning," *British Journal for the Philosophy of Science* 54 (2003): 553–567; Myrvold, W. C. "A Bayesian Account of the Virtue of Unification," *Philosophy of Science* 70 (2003): 399–423.

27. See Forster, M. and E. Sober (1994): "How to Tell when Simpler, More Unified, or Less Ad Hoc Theories will Provide More Accurate Predictions," *British Journal for the Philosophy of Science* 45: 1-35. For a reply, see Kukla, A. "Forster and Sober on the Curve-Fitting Problem," *The British Journal for the Philosophy of Science* 46, no. 2, (Jun. 1995).

28. Sober, E. "Two Uses of Unification." Pp. 205-216 in *The Vienna Circle and Logical Empiricism: Re-evaluation and Future Perspectives*, edited by Stadler, F. (Kluwer: The Netherlands, 2003), 214.

29. See Cartwright, N. *How the Laws of Physics Lie* (New York: Oxford University Press, 1983).

30. Cottingham, *Descartes* I 327-8.

31. Morrison, M., *Unifying Scientific Theories: Physical Concepts and Mathematical Structures* (Cambridge: Cambridge University Press, 2000) devotes chapter 1 to discussing "the many faces of unity."

# Chapter 4

# Lost Science, Deepwater Shipwrecks, and the Wheelbarrow of Archimedes

## Bridget Buxton

The nature of scientific innovation and its impact, in all societies including our own, is determined in significant measure by cultural factors. Cultural factors also play a role in the way we preserve and evaluate scientific achievement (or lack of achievement), which is invariably shaped by contemporary ideas about the purpose of science and the appropriate occupations of the scientist. In the study of Greco-Roman science, historical prejudices can be reinforced by modern assumptions, distorting not only the evidence that survives but the way in which it is interpreted. Modern perspectives on the foundations of western science must view the past through the filter of a centuries-old European intellectual establishment with an inbred distrust of revolutionary ideas and a widespread disdain for everyday mechanical innovation. Guided by these attitudes, the intellectual culture of early Christian Europe strongly favored the preservation of some texts over others, and our picture of the classical scientific achievement today remains limited by the focus of the ancient literary sources that survived the selection process.

To understand the true breadth of what Classical thinkers created and lost, particularly in the area of applied science, we must look beyond the texts of career intellectuals such as Aristotle and Ptolemy to chance references in a variety of sources, from imperial biographies to ancient building inventories. Archaeological discoveries on land are frequently supplementing this picture, but here too the evidence is limited to the artifacts and ecofacts that can physically survive several millennia of corrosion and decay. Such discoveries remind us of how little we actually know about the everyday technological innovations and scientific tools available to the ancients, but land excavation does not hold out much promise of dramatically improving the picture in the near future. The resulting situation is not unlike that envisioned in Plato's story of Atlantis: the ingenious mechanical works of the great Atlantean tinkerers sank to the bottom of the sea in a cataclysm of divine wrath, while the more abstract intellectual legacy of the city of philosophers (Athens) endured.

The lost technology theme in Plato's Atlantis myth may be truer than he knew. As underwater archaeology in the Mediterranean and Black Seas now enters a new phase of deep-sea exploration, there has never been a better opportunity to seek for lost technologies and even scientific texts from ancient shipwrecks. Total darkness, cold temperatures, minimal overburden, and (occasionally) lack of oxygen can contribute to excellent preservation conditions for organic materials at deepwater sites, provided they are spared the ravages of deep-sea trawling. At the same time, unfortunately, the budgets needed to conduct deep-sea excavations where such material might be preserved are far beyond the traditional funding options of the archaeologist, and still outside the main purview of the big oceanographic funding agencies. Yet the evolution of navigation and shipbuilding and study of the diffusion of such knowledge is surely of central concern in the history of science and worthy of scientific funding; finding the lost works of Archimedes and Eratosthenes would just be a bonus.

Deep submergence archaeology thus has the potential to contribute enormously to the study of scientific history in the pre-Christian Mediterranean, a globalized confluence of advanced and prosperous civilizations that in many ways resembles our twenty-first century world in its complicated attitudes to scientific innovation. Then as now, scientists could enjoy sponsored positions and relative academic freedom at the great libraries and courts of the ancient Mediterranean. The wealthy were free to pursue science as a hobby and many did so (Aristotle, Theophrastus, Lucretius, and Pliny the Elder, to name just a few). Social status was not necessarily a barrier to an aspiring intellectual, however. Consider the third century BCE mathematician Ctesibius of Alexandria, father of pneumatics, inventor of (among other things), a water organ, an improved water clock called the *clepsydra* that went into widespread use, and a fire engine equipped with a genuine double-action force pump; before gaining royal sponsorship he was apprenticed as a barber.[1] Yet then as now, scientific innovation was frequently channeled into paths dictated by cultural factors rather than available resources, or an awareness of how particular research might ultimately benefit or harm society.

# The Search for Hellenistic Science

The reason for targeting Hellenistic scientists as a case study is simple: the Hellenistic achievement in applied sciences surpassed that of the Romans and was in many instances not equaled until the modern era. First century CE Roman intellectuals such as Pliny the Elder were well aware that the Roman Empire was the beneficiary of this remarkable age of science. After admiring the works of his Greek predecessors on the subject of meteorology, which they accomplished during an era of turmoil and conflict, Pliny lamented that "now, during such a happy time of peace, under the rule of an emperor so pleased at the advance of literature and the arts, nothing whatsoever is being added to the sum of knowledge through original research, and in fact not even the discoveries made by our predecessors are carefully studied."[2]

To compare the Hellenistic achievement in applied sciences to our own remarkable age of innovation is, unfortunately, an unfashionable observation. According to a recent study of Hellenistic civilization, we should avoid indulging in "purposeless judgments and comparisons" and instead follow the approach of the pioneering scholar of Greek science, Geoffrey Lloyd, in focusing on the social context of scientific thought rather than its "achievements."[3] However, the concept of achievement has always been absolutely central to the way scientific innovation has been valued (or devalued) in western civilization, and therefore is inseparable from consideration of contemporary social context. Praise of achievement drove the competitive spirit that fuelled investment in intellectual resources from the time the earliest Greek tyrants sponsored a philosopher or collected a library, and had the equally important effect of ensuring that some scientists focused more on producing wonderful gadgets than practical tools.[4] The move to discard modern yardsticks of achievement that would place a steam engine above a spiked club, and accordingly deny the Hellenistic scientists their intellectual renaissance, seems to stem from a parallel urge to deny the existence of its opposite, the scientific stagnation that features in the cultural experience of both western and non-western societies. This uniquely modern way of thinking about (or rather avoiding thinking about) scientific achievement stands in stark contradiction to the avowed goal of its proponents to understand ancient science within an ancient context—since, as mentioned above, contemporary observers such as Pliny were quite comfortable with comparing Roman achievements in original research across cultures and centuries.

How, then, might we apply Pliny's rubric for evaluating ancient scientific achievement? It is clear that measuring the amount of research is meaningless unless there is another amount to compare it to, but still (according to one suggestion) we should try to avoid cross-cultural comparisons. "We should not assess the thinkers of this period on the basis of how far they fell short of what they might have achieved, or why they did not do the things they did not do."[5] This is a worthy caution against applying anachronistic expectations to ancient cultures, but unfortunately it may only serve to divert us from the very questions that make the study of ancient science of immediate value to the practice of modern science (and hence worthy of scientific funding). After all, when we ask why ancient science fell short of what we perceive to be its potential in specific instances, we are also revealing the cultural framework that governs the goals and expectations of our own scientific establishment. Moreover, as we shall see, the question of "what might have been" can sometimes lead to insights that improve not only our knowledge of ancient science itself, but our understanding of how that knowledge has been obscured and manipulated by cultural factors both ancient and modern.

# Wheelbarrows, Science, and Cultural Relativism

"Why should brains capable of conceiving a heliocentric universe, or of doing pioneer work on conic sections, so signally fail to tackle even the most elementary problems of productivity?"[6] The quote by the distinguished historian of Hellenistic antiquity Peter Green sums up what he felt was one of the great paradoxes of the Hellenistic Greek scientific achievement—here understood on its own terms as *historia*, a branch of philosophy involving experimentation and explanation of the natural world as well as technological innovation.[7] To put it in other words, as I once asked an undergraduate class in Hellenistic History, "if ancient Greek scientists were smart enough to work out the circumference of the Earth, describe circulation, and harness steam power, why didn't they invent the wheelbarrow?"

The spontaneous answers of these students were strikingly similar to the explanations found in modern articles and textbooks on the subject of Greco-Roman science, several of which make quite an issue of the troublesome absence of wheelbarrows. Perhaps social prejudice or religion was an inhibiting factor. Perhaps the preponderance of slave labor reduced the perceived need for labor-saving devices. Perhaps the ruling classes were terrified of devices that might lead to greater laziness and self-indulgence or, worse, more idle time for the working poor.[8] Perhaps the limitations of ancient sailing vessels prevented a more advanced understanding of ancient geography; perhaps the lack of telescopes limited the achievements of astronomy. The failure of many ancient inventions to "catch on" and have a wide impact was probably the result of cultural conservatism and skepticism about change; a reluctance to destabilize the status quo. In every case, interestingly, the students did not seem to be aware of the modern discovery that scientists can only deliver results in direct proportion to the amount of funding they receive. On the contrary, they invoked cultural factors to explain why some highly useful scientific ideas and innovations never appeared or failed to take hold while others, such as bronze casting, literally transformed the world.

The wheelbarrow exercise also demonstrates that any student with a passing familiarity with Greco-Roman civilization can readily grasp that the societies themselves, and certainly the ruling elites, were often constrained by their own cultural baggage to resist certain types of technological innovation. Typical history sourcebooks will feed students many examples that appear to support this conclusion. A favorite anecdote is the episode in chapter 18 of Suetonius' *Life of Vespasian* in which the otherwise notoriously practical emperor refused to use an ingenious mechanical device to help move some columns in order that more of the city poor might be employed for the task. To be fair, Vespasian also paid a reward to the engineer who made the suggestion; like most such stories used to illustrate anti-technological attitudes, there is another side.

It is easy to put our Classical forebears under the microscope and berate them for the cultural conservatism (or alternatively liberal "make work" projects) that seems to underlie their failure to achieve an industrial revolution when the scientific and technological know-how was clearly in place, possibly as early as the third century BCE. And yet in our modern obsession with science funding and marrying scientific progress to commercial objectives, do we really give enough thought to how our *own* scientific development (or in some cases, devolution) is culturally controlled? Again, my experience of a typical undergraduate ancient history class in the USA suggests not. I am leaving aside the parallel issue of widespread public ignorance of science itself (proven irrevocably by the advertising claims of the multi-billion dollar drug and beauty industries). The key point is that these students, who can typically come up with many good reasons why Plato did not invent a wheelbarrow, struggle to come up with even a single scientific or technological innovation that is in widespread use overseas but has yet to find a place in the United States. With prodding they might concede that the metric system is superior to inches and pounds, or allow that European and Asian cars are in many respects preferable to home-grown brands, or note that genetic engineering is subject to different legal limitations in different countries. They will even admit the role of corporations in steering the focus of medical, agricultural, and technological research, but invariably reject any consideration that scientific progress in the United States, the land of Edison and the Wright brothers, is *culturally* stifled. It takes a thorough immersion in the history of western science over the *longue durée* for them to start reflecting skeptically on the role of their own nation's cultural history in determining its scientific future.

Exposure to the classical historian's view of several millennia of western scientific renaissance and recession (and I will in this study be limiting the discussion primarily to Hellenistic science and the Classical World) should make it easier to understand—and perhaps confront—the ongoing impact of those elusive cultural stimuli and inhibitors, and to identify the reasons why one generation's Archimedes is employed by kings while the next is burned at the stake. Part of the value of trying to reconstruct the scientific thinking of our cultural ancestors is surely to improve our understanding of the more universal and enduring human reasons why science so often seems to fall short of its potential. We are not immune from history's cycles of ambitious scientific achievement and intellectual retreat, and we would do well to seek a greater understanding of the cultural factors steering the nature and impact of scientific innovation. This must start with a better knowledge of what our cultural ancestors actually found and lost, and why, before returning to consideration of those all-important social factors that Green understood to be at the heart of the failure of the Hellenistic Greeks to "even dream up so simple a device as the wheelbarrow."[9]

The interesting coda to this story is that Green and other modern scholars of ancient science may be entirely right about the major cultural impediments to technological innovation in the Greco-Roman world, but it turns out they are entirely wrong about the favorite iconic example, the humble wheelbarrow itself. We may now cite a little-known fifth century BCE epigraphic text recording building material inventories from the temple of Demeter and Kore at Eleusis in Mainland Greece that records a device called a one-wheeler (ὑπερτηρία μονοκύκλου) among other construction vehicles.[10] The document can only refer to a wheelbarrow. Subsequent indirect evidence for the use of this one-wheeled device appears until the fourth century CE in Greek and Roman sources. However, like so many ancient scientific and technological achievements, it disappears from the historical and material record in the early Christian centuries, only to reappear again in the High Middle Ages, sixteen centuries after it was first attested at Eleusis. Until quite recently, the existence of more ancient wheelbarrows was as inconceivable to classical scholars at the same time as it was, somehow, readily culturally explicable. This does not necessarily make the observations about Hellenistic scientists by modern historians such as Green incorrect, at least at a general level. It simply demands that we reconsider the nature of the ancient technological achievement, and reassess the limitations of the standard literary sources we traditionally rely upon to comprehend the ideals and scope of Greco-Roman science.

What do these literary sources really tell us about cultural attitudes to applied sciences in the Greco-Roman world? It is clear that in the main, perhaps even in cities as devoted to mechanical marvels as Hellenistic Alexandria, simple technical innovation, even to create scientific instruments, was considered a lesser calling than pure reasoning and the pursuit of abstract theory. We are told that Plato railed against philosophers who descended to petty mechanics and proposed that no resident citizen should practice the "technical crafts."[11] The Hellenistic inventor *par excellence* Archimedes reputedly refused to write a treatise on engineering because he did not want his great intellect to be associated with the fruits of banausic labor, according to a famous passage in Plutarch's *Life of Marcellus* written many centuries the inventor's death.[12] Elsewhere the Hellenistic mathematician Pappus of Alexandria noted that Archimedes did produce a treatise *On Spheres* concerning the construction of an orrery but did not consider it worthwhile to write about his other inventions.[13] The surviving writings of Archimedes do seem to reflect an interest in the theoretical over the applied, but does what survives of his writing truly reveal the attitudes of Archimedes and his contemporaries or merely the selective interests of later scholars and copyists? Would Archimedes' deficient publication record in the applied sciences have even attracted contemporary attention unless his choice was seen as unusual or inappropriate? And if Archimedes had invented a wheelbarrow, how would we know? The device speaks for itself: it does not require a treatise, but exists as its own explanation.

So, my titular wheelbarrow of Archimedes is of course a figment—Archimedes produced brilliant treatises and numerous practical innovations of scientific genius but he did not invent or write about the wheelbarrow because he did not have to. When he was born, it had already been in use for more than a century. The wheelbarrow myth accordingly points to an important and as yet insufficiently acknowledged weakness in our modern understanding of the practical achievement of ancient scientists. Namely, the very sources we rely on most heavily to understand this achievement, the literary sources produced by the elite landowning classes, are exactly the men we would least expect to take an interest in commenting on the marvels of the common wheelbarrow. One might expect such snobbery from philosophically inclined Greek aristocratic authors such as Plato and Plutarch, but even the Romans do not appear to present the more "practical" view they are usually given credit for.

Vitruvius, the great Roman architectural writer (who incidentally does not mention wheelbarrows either) clearly conceived of his treatise on the construction of public buildings as a marriage of theory and practice.[14] However, he also explained why his ideal architect must be well-versed in history, music, and moral philosophy as well as applied mathematics;[15] his science was both enabled and elevated by these humanistic foundations. The perceived "nobility" of his work helped to ensure its survival through the Middle Ages when many a text on matters of applied science and non-military engineering were lost through neglect, with the notable exception being agricultural manuals. Since the time of Hesiod and Cato, understanding such matters as bee-keeping was clearly thought to be appropriate for the land-owning elite; consider for example the revealing comments of the first century BCE Roman statesman Cicero: "of all the professions, none is better than agriculture, and none more profitable, enjoyable, or suitable for a free man."[16] The bias that has worked against the survival of information from ancient technical handbooks other than agricultural manuals goes back at least as far as Aristotle, who clearly had access to such works but did not reproduce their content.[17]

The Classicist K. D. White, one of the leading scholars of Greco-Roman technology and agriculture of the last century, tackled this issue of ancient and modern attitudes provocatively in an essay entitled "The Base Mechanic Arts?"[18] He also drew connections between modern academic snobbery and alienation from banausic crafts and the assumed prejudices of the ancients, with the former reinforcing the latter to distort our view of the Classical achievement. Yet these prejudices, ancient as well as modern (White says "universal") cannot have been all-pervasive. It is just that over time, and combined with the forces of decay and destruction that have obliterated so many of the treatises and technologies themselves, they have exercised a brutal censorship over our access to the Greco-Roman scientific achievement, particularly in the areas of applied science and technology. So little remains, it is inevitable that references to ancient science in our familiar sources such as Plutarch now assume an artificial and perhaps distorting prominence.

To illustrate this problem, White takes the example of Plutarch's comment about Archimedes' disdain for his own inventions and reported failure to publish on matters of applied science, and puts it in its proper historical context. Plutarch was no engineer, he points out, and everything we know about Archimedes' role in the defense of Syracuse and his complete transformation of Egyptian agriculture through the introduction of the water-screw that still bears his name argues against the idea that Archimedes or his sponsors devalued technological innovation. Probably the invention of steam-powered canons, reportedly part of the arsenal of Syracuse, was less impressive to Archimedes than his equation for determining the number of grains of sand in the universe (on which he did write a surviving treatise). However, that does not mean that he held the former achievement in contempt. Plutarch, living comfortably under the *pax Romana* and immersed in the gentlemanly philosophies of the second sophistic, might have held such attitudes, but it is hard to credit them in the man who claimed that his mechanical inventions could move the earth. This is a boast that Archimedes followed up far more effectively than any publication by offering a practical demonstration of a crank and pulley system that could move a full-sized merchantman.[19] The device itself was worth a thousand words.

# Unbreakable Glass and Invisible Gadgets

To glimpse the technical achievements of the Hellenistic scientists, we often need to read between the lines. Interestingly, modern researchers from the time of Galileo attempting to recreate Archimedes' 'Eureka!' discovery of the technique for determining the purity of gold in the crown of King Hiero II have concluded that his method is only explicable if he also developed a measuring device (actually just a simple scale) to exploit the principle of displacement to *make* it work. The differences in displacement between the adulterated and pure gold would otherwise have been too minute to observe by simple immersion of the crown. If Archimedes employed the necessary scale to compare the crown's buoyancy with the equivalent weight of pure gold, as seems extremely likely, to solve the problem of the crown, then it is very telling that we do not hear about it in surviving versions of the story—which incidentally does not appear in any known writings by Archimedes himself. However, the understanding of hydrostatics necessary to create an appropriate scale to measure the buoyancy of the crown is clearly demonstrated in his surviving treatise *On Floating Bodies*. Later writers who passed on the crown story, such as Vitruvius, were interested in Archimedes as a celebrity, and the Eureka story with its memorable leap from the bathtub and naked run through the town probably endured because it gave appealing color to his idiosyncratic personality. The exact nature of the scientific discovery illustrated by Archimedes' bathtub experience (if it ever occurred) and the device actually needed to test Archimedes' theory were incidental to the literary goals of the texts in which the crown story later appeared.

Occasionally and indirectly, we are given a glimpse of other ancient scientific innovations that have now disappeared from the historical and material record. The most memorable story concerns the emperor Tiberius Caesar in the early first century CE, and slightly different versions of it are recorded in the encyclopedia of Pliny the Elder and his contemporary Petronius.[20] One day an unnamed inventor devised a method for making a goblet of flexible and unbreakable glass, which when dashed to the floor could be returned to its original shape with a bit of delicate hammering or manipulation. When he presented this discovery to Tiberius, hoping for a great reward, he was asked if anyone else knew its secret, and answered no. The emperor immediately had the unlucky inventor beheaded and his workshop destroyed, to prevent the new product from undermining the value of gold and silver. Cassius Dio, a second century CE Greek historian, repeats the story and adds that the inventor was an architect who had already fallen out with the emperor on account of the latter's jealousy of his technological achievements.[21] One is reminded of the previously mentioned story about Vespasian refusing to adopt a labor-saving technology for the sake of a larger economic good, begging the question of whether we are dealing with some kind of literary trope rather than a biographical fact. Can the story be trusted?

When we consider that ancient Roman glass was typically colorful to the point of opaque, it is not impossible that the "flexible glass" in question was actually aluminum (it is very unlikely to have been plastic as some have proposed). The existence of aluminum (aluminium) was not formally (re)established by western scientists until the early nineteenth century, and effective methods for extracting the pure metal did not appear for some decades later in Western Europe. That such a discovery was not beyond the capabilities of the ancient Romans is surely proven by their works of metal and glass that do survive. The famous Portland vase from the time of Augustus took the eighteenth century potter Josiah Wedgwood four years to duplicate in jasperware, and in the nineteenth century the competition to duplicate the vase in glass was won by glassmaker Philip Pargeter, a project that took him three years.[22]

Even the exquisite Portland vase seems unsophisticated in comparison to works such as the Constable-Maxwell cage cup, an extremely rare *diatretrum* or Roman glass cage cup from the third century CE.[23] The existence of such incredible artworks and the knowledge required for their making is something that is attested only by the half-dozen known surviving cage cups themselves. Significantly, the craft may have reached its apogee in Europe in the third and fourth centuries CE, centuries of economic and political collapse in the Roman Empire. But the greater issue and question concerns what other glass instruments have been lost in the interim. One of the explanations frequently given for the limited progress of some ancient scientific disciplines is a lack of precise scientific instruments for measuring and observation.[24] However, in this case the absence of evidence is not necessarily evidence of absence: we should keep in mind the fate of the unsung gadget of Archimedes' Eureka story. Also, the paucity of ancient glass, wood, and metal objects that have survived from antiquity is mainly due to their fragility and (in the case of metal) recyclability.

If the literary record is skewed and incomplete to the point of untrustworthy, what other recourse do we have to improve our understanding of the lost achievements of Classical science? The archaeological record is clearly the major untapped resource, even if its yield is erratic and unpredictable.[25] And yet it is always coming up with surprises (like references to wheelbarrows) that force us to rethink our view of the impractical ancients and return to the troubling question of why useful discoveries and innovations can sometimes disappear as readily as they appear. One such recent discovery obtained by the Fitzwilliam museum in 1991 has been dubbed as "the Roman Swiss army knife."[26] Anticipating its modern counterpart by over eighteen centuries, it was silver apart from a blade of iron, and featured a selection of fold-out picks and spikes as well as a fork, a spoon, and a spatula.[27] Such surprising finds are the exceptions that prove the rule: they simply remind us that we have no real idea about how much we have lost. Clearly over several millennia, the archaeological record can be just as drastic a censor of ancient technology as the literary record.

However, there is one place where the ancient material record can some-times be preserved almost untouched in the most neatly packaged time-capsule, and all estimations suggest that thousands of such time-capsules are still await-ing discovery in the dark waters and anoxic mud of the deep sea. The challenge to enter this new frontier was first raised by oceanographer W. Bascom with his influential 1976 book entitled *Deep Water, Ancient Ships: Treasure Vault of the Mediterranean*. Bascom used insurance statistics from the age of sail to calcu-late that a very large number of ancient shipwrecks (though probably still only 20 percent of the total) must have occurred in the deep water of the open sea. The archaeological potential of such deepwater sites has more recently been dis-cussed by underwater archaeologist S. Wachsmann, one of several American ar-chaeologists who have partnered with ocean scientists to find and investigate ancient shipwrecks in the deep water of the Mediterranean and Black Seas.[28] Oceanographer R. Ballard has been the leader and driving force behind many of these projects, and his engineering team has played a critical role in the devel-opment of deepwater excavation techniques in the United States. Other trail-blazers too could be mentioned, but none have been so forthright as Ballard in challenging the scientific community to support deep submergence archaeology with oceanographic-scale funding: "Is the study of human history beneath the sea just as important as the study of natural history beneath the sea? Are cultural resources as significant as natural resources?"[29]

These questions reflect a contemporary cultural myth that culture and sci-ence are somehow separate and unequal concerns, a myth that is perpetuated by the unequal funding regimes that support the study of natural and cultural re-sources. To the historian, however, the reality is self-evident: science itself *is* a cultural resource, and scientific achievement is fundamentally a cultural phe-nomenon. The evolutionary path of science within Western Civilization, along with all the cultural baggage that we have kept along the way, will surely con-tinue to play an important role in shaping our scientific future. Whether or not our scientific achievement will appear to future generations to have reached its apparent potential will be influenced by a myriad of cultural factors, the effects of which may be greater if their negative potential is underestimated. By shaping the research environment, cultural factors continue to influence scientific achievement as much as they ever did, and history offers us many demonstra-tions of how the former can be fatal to the latter.

Clearly the study of scientific history is worthy of scientific resources. The more pertinent question in this case, however, is whether the investigation of an-cient shipwrecks (and in particular deepwater shipwrecks) represents a good in-vestment of those resources compared to other avenues of research. Let us con-sider for a moment what a shipwreck represents. A shipwreck is a snapshot of the world on the day of its creation, and a ship is an unbiased and indiscriminate collector of technology, as well as an important scientific achievement in itself. Praising the invention of ships in his *Natural History*, Pliny the Elder asked: "for what is more violent than sea, winds, whirlwinds, and storms? By what greater invention of man has Nature been aided in any part of herself than by sails and oars?"[30] There were many marvelous ships constructed in the Classical world, from the rapier-thin trireme battleships of the Athenians to great merchantmen of the Veneti of Brittany (Julius Caesar's description of these vessels suggests they might easily have handled a transatlantic voyage, and contemporaries certainly possessed the geographical knowledge to justify the attempt).[31] *On the Ship*, the second century CE writer Lucian's account of a Roman grain ship that appeared unexpectedly in Piraeus, depicts residents of Athens amazed by the vast size of the vessel, which could carry enough grain to feed all of Attica for a year. We know of Roman cargo ships of over 1,200 tons burden; merchant fleets of this size did not appear again until the British East Indiamen at end of the eighteenth and beginning of the nineteenth centuries.[32] And yet interestingly, there was no consistent progression in size and sophistication as ship-building evolved from reliance on sewn hulls and mortise-and-tenon shell-first construction to the modern technique of frame-first construction with copper and iron nails. The brilliant variety of ancient ship designs in itself serves as a fascinating multi-regional case study in technological evolution.[33]

There is every reason to expect that the contents of the grander vessels of the Hellenistic world and Roman Empire would have been as impressive as their construction. We are told of Hellenistic ships equipped with banquet halls, gardens with automatic watering systems, a library, fish ponds, and even stables.[34] Such vessels would seem unbelievable but for the discovery of the Roman emperor Caligula's huge floating pleasure palaces sunk in Lake Nemi.[35] Ancient shipwrecks have also preserved unique collections of everyday practical objects. The largest collection of Byzantine wood-and-iron tools and by far the largest collection of Medieval Islamic glassware have both come to us from shipwrecks.[36] When one considers that virtually everything that humans have ever made or written must at some point have travelled by sea, and that it is estimated that the life of half all ships ends in shipwreck[37]—the only real questions concern where such material is likely to be best-preserved, and how soon can we get down there? A veritable Atlantis of lost science awaits discovery on and under the sea floor. In fact, we have known about the Mediterranean Sea's potential for revealing lost technology ever since the earliest days of underwater archaeology in Greece.

## Technology From Deep Water

Sometime in 1900, a storm drove sponge diver Captain Dimitrios Kondos to seek shelter on the northeastern shore of the small Greek island of Antikythera. As his ship waited at a place where the cliffs fall steeply into the sea, Captain Kondos sent a diver down to look for sponges. But the diver had hardly touched bottom when he gave the "come up" signal. When they unscrewed his helmet, the man was babbling, half screaming, something about writhing horses, naked women, and syphilitic corpses. Captain Kondos went down to have a look for himself, and in time returned to his ship carrying a man's right arm, in life-size bronze, fingers extended as if to hold an object. Later, the rest of the statue would be retrieved, and is now one of the centerpieces in the hall of Classical bronzes in the National Museum in Athens. It is one of dozens of spectacular Classical Greek statues recovered from what was likely a Roman shipwreck of the early first century BCE.[38]

The salvage of the Antikythera treasures was really the forerunner of the first systematic underwater excavation—but it came at a heavy price. The site itself was clumsily destroyed, and several divers died of the bends. Another tragedy came to light after the excavators—who were all sponge divers rather than archaeologists—had rolled a number of large boulders off the site and down into the unreachable depths. When one of these so-called boulders was raised to the surface it was found to contain a bronze statue.

Certainly there have only been a few ancient shipwrecks as rich as Antikythera, but its greatest treasure was the fragmentary and incomplete remains of a 2000-year-old ancient machine we now call the Antikythera mechanism, an incredibly complex astronomical device that has been described as a forerunner of the modern computer. The recovery and reconstruction of the mechanism's interlocking bronze gears from eighty-two corroded fragments continues to this day, and has been a cumulative process, with important contributions by the late historian D. De Solla Price, who claimed that the device "requires us to completely rethink our attitudes toward ancient Greek technology."[39] The device made science headlines again with an article by T. Freeth, A. Jones., J.M. Steele, and Y. Bitsakis in the journal *Nature* revealing further possible functions of the mechanism.[40] The abstract of that article (slightly abridged) gives some indication of the complexity of the mechanism and efforts to decipher its function:

Previous research on the Antikythera Mechanism established a highly complex ancient Greek geared mechanism with front and back output dials. The upper back dial is a 19-year calendar, based on the *Metonic* cycle, arranged as a five-turn spiral. The lower back dial is a Saros eclipse-prediction dial, arranged as a four-turn spiral of 223 lunar months, with *glyphs* indicating eclipse predictions. Here we add new and surprising findings concerning these *Back Dials*. Though no month names on the Metonic calendar were previously known, we have now identified all twelve months, which are unexpectedly of Corinthian origin. The Corinthian colonies of northwestern Greece or Syracuse in Sicily are leading contenders—the latter suggesting a heritage going back to Archimedes. Calendars with *excluded days* to regulate month lengths, described in a first century BC source, have hitherto been dismissed as implausible. We demonstrate their existence in the Antikythera calendar and in the process establish why the Metonic Dial has five turns. The upper subsidiary dial is not a 76-year Callippic Dial as previously thought but follows the four-year cycle of the *Olympiad* and its associated *Panhellenic Games*. . . . We explain the four turns of the Saros Dial in terms of the *Full Moon Cycle* and the lower subsidiary *Exeligmos Dial* as indicating a necessary correction to the eclipse times in the glyphs. The new results reveal an unforeseen provenance and additional sophistication in the Antikythera Mechanism. The Metonic Calendar, Olympiad Dial and eclipse prediction link the cycles of human institutions with the celestial cycles embedded in the Mechanism's gearwork: a microcosm of the temporal harmonization of human and divine order in the Classical world.

The Antikythera mechanism surely deserves to displace the wheelbarrow as the poster-child gadget that the ancient Greeks, by most modern historical reasoning, never should have developed. It seems curiously appropriate that such an extraordinarily high-tech device should be devoted to the most cultural of concerns, if calculating the dates of the great Panhellenic athletic festivals was truly one of its primary functions. If this interpretation is correct, it serves as an important reminder of that fact that, in the words of L. White, "no Greek or Roman ever tells us, either in words, or in iconography, what he, or his society, wanted from technology, or why he wanted it."[41]

The link to a Corinthian colony revealed by the research of Freeth *et al.* is also interesting, because it puts us in the territory of Archimedes, whose ingenuity may well lie behind the science of the mechanism. There even appears to be a description of a similar device in the writings of the Alexandrian astronomer Ptolemy and the first century Roman statesman Cicero, who attributes it to Archimedes—but who could have believed him if such a thing had never been found?

> [W]e have learned to survey the stars, not only those that are fixed, but also those which are improperly called wandering; and the man who has acquainted himself with all their revolutions and motions is fairly considered to have a soul resembling the soul of that Being who has created those stars in the heavens: for when Archimedes described in a sphere the motions of the moon, sun, and five planets, he did the very same thing as Plato's God, in his Timaeus, who made the world, causing one revolution to adjust motions differing as much as possible in their slowness and velocity. Now, allowing that what we see in the world could not be effected without a God, Archimedes could not have imitated the same motions in his sphere without a divine soul."[42]

Whether Cicero is referring to something similar to the Antikythera device or perhaps a full-sized orrery / planetarium we cannot know, but we should note his clear admiration for the inventor as well as the achievement. The survival of Alexandrian texts describing the marvelous devices of men like Heron, Philo, and Ctesibius suggest that mechanical marvels from the frivolous to the deadly were celebrated and sought after in the courts of Hellenistic rulers, an interest they most likely passed on to the Romans. But the concerns of Hellenistic engineers also extended into a few everyday items. The Archimedean screw we have already encountered; the water-clock of Ctesibius was in widespread use.[43] Ancient generals quickly grasped the military applications of the torsion catapults first developed in Syracuse, but the same technology also appeared in the runner's starting gate at the athletic stadium in Nemea.[44] These devices present no challenge to our assumptions about ancient science because they have come to us partly via technical manuals that provide enough information to reconstruct working models. We can understand the mechanics of Heron's self-opening temple doors,[45] and may even credit the fourth century philosopher Archytas with the creation of a mechanical flying bird;[46] mechanical singing birds were already a commonplace.[47] All such things fit common assumptions about the interests and capabilities of the Hellenistic scientist: a lover of marvelously complex and unique gadgets rather than practical mass-produced innovations.[48] But where do we draw the line between descriptions of gadgetry that we can comprehend and build ourselves, and ancient inventions that seem to belong more to the realm of science fiction than credible history?

Gadgets that defy belief may be found in texts as old as Homer, who describes the divine construction of a fleet of wheeled golden tripods that would travel of their own accord to and from the councils of the gods—the original "meals on wheels."[49] Greek, Roman, and early Arabic literature is filled with stories of lifelike statues brought to life through the genius of their creators, ranging from beautiful women[50] to a self-powered mechanical snail that even left a trail of slime.[51] Where to draw the line? No one has questioned the report of Appian that Julius Caesar's funeral saw the presentation of a mechanical wax manikin of the dictator that would turn to display its wounds to the crowd.[52] But what can be made of the mechanical monsters encountered by ancient travelers from Odysseus to Sinbad the Sailor, whose real-world Arabia knew of a myriad of complex automata frolicking in the pleasure gardens of early medieval Caliphs?[53] Maybe nothing, except in this case we actually still have the written instructions for creating such robots by the twelfth century Arabic inventor Abu al-Iz Ibn Ismail ibn al-Razaz al-Jazari, author of the *Kitab fi ma'rifat al-hiyal al-handasiyya* (Book of Knowledge of Ingenious Mechanical Devices) published in 1206.[54] Significantly, Al-Jazari (as he is usually called) must have had access to many of works of his Alexandrian predecessors, including Heron, some of whose works survived only in Arabic. The Arab genius may also have inspired the Italian Renaissance inventor Leonardo da Vinci, whose design for a mechanical knight bears a striking resemblance to a similar automaton known to Al-Jazari. There may even be a continuous intellectual evolution from the wings of Icarus to the flying machines of Da Vinci via the texts of Hellenistic, Roman, and later Arabic scientists. If so, certain factors common to all pre-industrial societies probably kept these remarkable automata and the instructions for making them in very limited circulation.

This brings us back to the question of whether the deep oceans might really represent an Atlantis of undiscovered ancient miracle gadgets—for certainly at this present time, the Antikythera mechanism stands alone. The ideal anoxic conditions for preserving some otherwise perishable organic materials in the water column are currently only found in the depths of the Black Sea below 150m, which is somewhat peripheral to the most well-travelled routes along which valuable items were exchanged in the pre-Christian era.[55] In the Mediterranean, the best case survival scenario for organic and metallic remains in an ancient shipwreck is for a section of the hull to become buried in the anoxic seafloor mud by the weight of the cargo above. Reaching these deeper layers demands excavation, however, which in the deep-sea is a difficult and expensive undertaking, especially when the archaeological payoff is never guaranteed. For the conditions to be right for the discovery of another Antikythera mechanism, archaeologists will probably have to wait for the equivalent of another Antikythera shipwreck—one so obviously laden with bronzes that the necessity of excavation is not in doubt. There must be many such ships still to be found in the Mediterranean, as we have already come across evidence or the actual remains of approximately a dozen—unfortunately not before they were torn apart by trawlers and looters.[56]

# Lost Libraries

The odds of coming across an Atlantis-sized haul of lost ancient science and technology improve markedly, however, if we keep in mind that a vast amount of ancient scientific thought was at some point recorded in books. Even if some of their works are known only by title, the output of Heron and Ctesibius of Alexandria, Philo of Byzantium, and even Al-Jazari suggest that publication of technological treatises was both normal and expected, even for a busy polymath like Archimedes whose day jobs included defending Syracuse from Romans. Such works would have found their place among a very broad and long-lived scientific literature, from the farming manuals of Cato to the geography of Strabo, the encyclopedia of Pliny to the medical writings of Galen, the architecture of Vitruvius to the emperor Domitian's lost manual on hair care.[57] These manuscripts would in turn have been duplicated in the great libraries of Rome, Pergamum, and Alexandria among thousands of other texts dealing with natural and applied science, including such gems as the Pseudo-Aristotelian treatise on the nature of toenails, and Theophrastus' essay "On Smells." Then as now, the quality and relevance of academic publishing was greatly variable.

The highest achievements of Classical civilization were achievements of the mind, and the discovery of ancient libraries—as the primary repositories of all this collected philosophy, science, and literature—represents one of the ultimate holy grails of classical archaeology. Books, however, might seem to be the most fragile and unlikely objects to seek under the sea. The survival of a few Greek texts in Arab collections notwithstanding, it is well known that the great library of Alexandria was utterly destroyed: by Julius Caesar in the first century, Aurelian in the third century, Theophilus in the fourth century, and Caliph Umar in the seventh century.[58] There are problems with all these stories, but in Julius Caesar's case Michael Grant has suggested that the main loss of books were of those waiting in storage at the docks, where many warehouses were burned.[59] The bulk of the great library clearly survived the Alexandrian war, and continued to flourish. We do not know if it also continued to take copies of all new books found on arriving ships, a practice begun by order of the Ptolemies that we know dates back to the early third century BCE.[60] Regardless, it is clear that Egypt, the only place in the ancient world that produced significant amounts papyrus paper, remained the hub of a major maritime book trade for at least half a millennium. Books were traded by sea as readily as any other common or valuable commodity in the ancient world.

Clearly it is one thing to know that ancient ships carried books, and for that matter letters: a vast personal and bureaucratic correspondence is attested even in the Bronze Age. It is quite another to find them intact and readable. But there are some places in the deep ocean where even a sealed leather-bound scroll, a late Roman codex, or a medieval palimpsest could survive. The leather cordage at the top of the mast of the Sinop D Byzantine shipwreck off the Black Sea coast of Turkey still survives after fifteen centuries of exposure in the water column at 320 meters.[61] And it does not even require such an extreme environment for treated leather to achieve this level of preservation. More than two miles under the Atlantic, in the oxygenated waters next to the ruin of Titanic, benthic organisms have devoured all trace of human remains. However, leather shoes and suitcases rest in the sand as if they had been deposited yesterday. Could not vellum do the same? In these examples, too, we are looking only at scenarios where the material is fully exposed in the water column. Large ships (the kind most likely to favor open-water routes and carry small libraries) would have plunged deep into the sediment upon sinking, sealing off lower parts of the hull and creating an environment suitable for the preservation of organic materials. And even not-so-large ships in not-so-deep water can contain surprises. The oldest known book in the world, a wooden diptych from the 1400 BCE Uluburun shipwreck discovered off the coast of Turkey, survived 34 centuries inside a ship's cargo jar that had once contained pomegranates.[62]

The best possibilities for discovering ancient books and other organic artifacts however lie in deeper water. Since Robert Ballard's discovery of RMS Titanic using remotely operated vehicles in the 1980s, the science and technology of finding shipwrecks in deep water has gone from strength to strength. Still, a lot of factors are working against an archaeologist hoping to succeed in this line of research. Costs remain prohibitive for humanities-sized archaeological budgets, leaving oceanographic teams and well-funded private groups to dominate the field and set the priorities. Deeper wrecks that may have escaped the attention of sports divers and sponge-draggers are frequently obliterated by deep-sea trawlers, which drag nets held down by giant weights across the sea floor. The effect of these trawling nets on an ancient shipwreck is comparable to a museum being torn apart by a bulldozer. And even when a well-preserved deepwater shipwreck is found, what then? Tools and techniques available for remote excavation are still imperfect, and the nature of the available funding as well as political issues invariably limit the amount of sampling and investigation that can take place.

The main good news is that with the help of experience and new technology, we are getting noticeably better at finding and recording deepwater shipwrecks every year. This development has certainly been helped by NOAA (the U.S. National Oceanographic and Atmospheric Administration) creating new funding opportunities for ocean exploration, funding which allows limited forms of archaeological survey to piggy-back on science cruises devoted to more conventional oceanographic research. With NOAA support, a team from the University of Rhode Island has been imaging the seabed below 80m off the western coast of Turkey from 2008-2010, first using only a small dive boat and an Outland ROV (remotely operated vehicle). In 2009 we continued this underwater imaging project with the 65m research vessel Nautilus and the full suite of R. Ballard's Institute for Exploration ROVs. During these three short cruises, we have averaged the discovery and investigation of at least one new shipwreck for approximately every twelve hours of search time. With the right tools, research, and experience, finding ancient shipwrecks in deep water is now becoming a bit less like finding a needle in a haystack and more like finding a needle in a hay bale. The vast majority of these discoveries are not exceptional, and none to date are likely destined for more than surface mapping and *in situ* preservation. What will happen, then, when we eventually find the next Antikythera shipwreck, or a gigantic Roman freighter, or Hellenistic battleship in pristine condition on the sea floor?

Wachsmann, who has given extensive consideration to the problems of deep-sea excavation by remotely operated vehicles, does not think that securing funding for such an excavation will be a problem.[63] If the shipwreck is obviously important enough, governments and individuals will very likely rise to the challenge. When this happens, it will undoubtedly be for different reasons than the untested possibility of finding the vellum or leather-bound documents we know that many ships undoubtedly carried. On this list we may included the captain's charts and *periploi (*sailing guides), novels and poetry, or even a long-lost scientific library. The latter possibility is really not as remote as it might sound, since we know that important scientific and philosophical works were hoarded, traded, and coveted in antiquity. Many would have understood the Roman general Sulla's desire to obtain the library of Apellicon of Teos, which included most of the works of Aristotle and Theophrastus.[64] The only known large private library that survives from classical antiquity, the charred collection of scrolls from the villa of the Papyri in Herculaneum, appears so far to be philosophical in focus.[65] Finding a comparable collection of long-lost texts on a sunken ship is really no more unlikely than finding another cache of Hellenistic bronzes, and would surely rank as underwater archaeology's greatest "Eureka!" moment.

# Conclusion

So—to summarize, we know what is probably down there. We know that for the last four thousand years at least, there has been a flourishing seabourne trade across deep water. Secondly, we know about the where. Given the time and re-sources, we know where and how to find these ancient ships, and we are contin-ually developing better ways to record and excavate their contents. Finally, and most importantly, we know about the why. There are many whys, in fact, but surely one of the most important archaeological virtues of an ancient shipwreck is that it is a true time-capsule, offering us glimpses of the Classical scientific and technological achievement that can hardly be rivaled by anything on land outside of Pompeii and Herculaneum. It is nothing short of amazing that, given recent technological and methodological advances in underwater archaeology, and the repeated appearance of the Antikythera device in the world's foremost science journals, that the excavation of the Antikythera shipwreck has yet to be completed. More fragments of that remarkable device—and perhaps even others like it—are still buried in the sand of an excavation that was abandoned over a century ago. As might be expected, the reasons are complex, but lack of adequate scientific funding continues to be a major problem for deep-sea archaeology.

In conclusion, archaeologists and scientists alike should be optimistic that future discoveries will help to demonstrate what ocean explorers already know to be true: that the deep ocean is the world's greatest museum of human history, and its investigation could contribute enormously to our understanding of the achievements and evolution of western science. Gaining a greater knowledge of the Classical scientific achievement and understanding why this remarkable age of innovation and discovery gave way to a self-imposed scientific Dark Age may offer lessons for our own times that transcend the backward-looking interests of archaeologists. By using archaeology to gain a more accurate picture of what Classical scientists achieved and Christian Europe subsequently lost, we may even come closer to a cultural understanding of why our own scientific progress is seldom linear, always surprising, rarely inevitable, and apparently, never permanent.

# Notes

1. Ctesibius wrote a number of scientific treatises but none survive. On his inventions see Vitruvius, *On Architecture* 10.7.1-4 and 9.8.2-7 (his early life in his father's barber shop); Philo, *Pneumatics* app. 1, chap. 2.

2. Pliny the Elder, *Natural History* 2.117-118, translation from Humphrey *et al.* 1998: 591.

3. Shipley, G. *The Greek World After Alexander 323-30 BC* (London: Routledge 2000), 326.

4. A trait for which the Alexandrian court inventors are frequently singled out for disapproval; see Green, P., *Alexander to Actium: The Historical Evolution of the Hellenistic Age* (Berkeley: University of California Press, 1990), 479.

5. Shipley, *Greek World*. 2000: 327.

6. Green, P., *Alexander to Actium: The Historical Evolution of the Hellenistic Age* (Berkeley: University of California Press, 1990), 467.

7. White, K.D., "The Base Mechanic Arts? Some Thoughts on the Contribution of Science (Pure and Applied) to the Culture of the Hellenistic Age," in P. Green (ed.) *Hellenistic History and Culture* (Berkeley: University of California Press, 1996), 211.

8. An idea also explored in Green, *Alexander,* 469.

9. Green. *Alexander:* 474.

10. For the definitive study of the ancient wheelbarrow, see Lewis 1994: 453-475.

11. Plutarch. *Moralia* 8.2.718e; Plato. *Laws* 8.84d-847a.

12. Plutarch. *Life of Marcellus* 17. Plutarch was a prolific author with philosophical interests who held the privileged status of a priest of Delphi during the second century CE, and is our main source for the story of Archimedes' almost single-handed defense Syracuse from the besieging Roman armies of Marcellus in 214-212 BCE. Plutarch was concerned to contrast Archimedes' reputed dislike for using his great intellect to minister to "everyday needs" with his enormously influential innovations in applied science, particularly in the area of military technology. The passage is frequently cited to illustrate the existence of social prejudices against the writing of technical handbooks; see Humphrey *et al.* 1998: 583-584.

13. Pappus of Alexandria. *Mathematical Collection* 8.3.

14. Vitruvius. *On Architecture* 1.1-4.

15. Vitruvius. *On Architicture* 1.5-8.

16. Cicero. *On Duties* 1.151.

17. Humphrey *et al.* 1998: 583.

18. White, "Base Mechanic Arts," 211-220.

19. Plutarch. *Life of Marcellus* 14.7-9.

20. Pliny the Elder. *Natural History* 36.66; Petronius. *Satyricon* 51.

21. Dio 57.21.7. For a brief discussion of variations of the story see Lazenby, F.D. "A Note on Vitrum Flexile," *The Classical Weekly* 44, no. 7 (Jan. 15, 1951): 102-103.

22. Keynes, M. "The Portland Vase: Sir William Hamilton, Josiah Wedgwood and the Darwins," *Notes and Records of the Royal Society of London* 52, no. 2 (Jul., 1998): 237-259; Philip Pargeter's imitation may now be found in the Corning Museum of Glass in New York.

23. An auction house description of the cup may be found at http://www.bonhams.com (accessed July 10, 2010). In 2004 the cup was sold at auction to an unknown buyer for the highest price ever paid for a single piece of glass: £2,646,650 ($4.8 million).

24. Green, *Alexander*, 457; however see also 850 n. 30, where he suggests the existence of sophisticated instruments of observation to support ancient navigation is quite likely, but in the main outside the interests and surviving writings of the major scientific theorists. On ancient navigation instruments however see now Davis 2009, who demonstrates that Greek and Roman open water navigation techniques did not necessarily require sophisticated ship-board instrumentation.

25. A useful introductory comment on the value of archaeology in reconstructing the history of technology may be found in Humphrey, J.W., Oleson, J.P., and Sherwood, A.N. *Greek and Roman Technology: A Sourcebook: Annotated Translations of Greek and Latin Texts and Documents* (New York: Routledge, 1998), xii-xxiii.

26. Fitzwilliam museum. Accession Number GR.1.1991.

27. Vassilika, E., *Greek and Roman Art* (London: Cambridge University Press: 1998), 128-129, no. 62.

28. See Wachsmann, S. "Deep Submergence Archaeology: the Final Frontier," *Skyllis* (Deutsche Gesellschaft zur Förderung der Unterwasserarchäeologie e.V.) Proceedings of *In Poseidon's Reich* XII.8.1-2 (2007-2008), 130-154; also see Ballard, R. (ed.) *Archaeological Oceanography* (Princeton: Princeton University Press 2008) for American perspectives on the state of the field of deep submergence archaeology / archaeological oceanography. The work of Anna Marguerite-McCann and colleagues at Skerki bank, Cheryl Ward's foundational work on the Sinop D shipwreck, and Laurence Stager's investigation of the Ashkelon shipwrecks all took place in partnership with oceanographer Ballard's Institute for Exploration. Conservator Dennis Piechota has also been a long-term collaborator with Ballard and has made significant contributions to the study of artifact preservation in deep sea and Black Sea environments (see Piechota, D., Ballard, R., Buxton, B., Brennan, M., "In situ Preservation of a Deep-Sea Wreck Site: Sinop D in the Black Sea," (forthcoming in the proceedings of the ICC 2010 Istanbul Conference: Conservation and the Eastern Mediterranean)). The author has participated in IFE expeditions in the Black Sea and Aegean led by Ballard since 2007.

29. Ballard, Robert. *Archaeological Oceanography* (2008: 1).

30. Pliny the Elder. *Natural History* 32.1.

31. On the wonders of the trireme (one of which is that we have not yet resolved all the issues of its construction), see Hale 2009. Julius Caesar's sea battle with the Veneti is recounted in book three of his *Gallic War*. Strabo in his *Geography* (2.3.6) anticipated

Columbus and probably followed his Hellenistic predecessors in speculating about reaching India across the Atlantic. The first century CE Roman philosopher and statesman Seneca anticipated a future age of exploration that would reveal new lands across the sea (*Medea* 375). It was, of course, widely and firmly established in the ancient world that the earth (*orbis terrarum*) was a globe.

32. Casson, L. *Ships and Seamanship in the Ancient World* (Princeton: Princeton University Press, 1971), 188.

33. The classic and still unsurpassed study is Lionel Casson's 1971 *Ships and Seamanship in the Ancient World*. A general overview of Institute for Nautical Archaeology shipwreck excavations and their contribution to our understanding of ancient Mediterranean ship design may be found in Bass 2005. The pioneering work on this subject was J. Richard Steffy's 1994 *Wooden Ship Building and the Interpretation of Shipwrecks*, and a more complicated view of the evolution of ancient ship construction is still unfolding, with numerous interesting discoveries now coming out of Israel (see in particular the publications of Y. Kahanov and colleagues at the Recanati Institute for Maritime Studies, http://hcc.haifa.ac.il/Departments/maritime/staff/ykahanov.html (accessed July 10, 2010)).

34. Athenaeus. *Philosophers at Dinner* 5. 203-204; 206-209.

35. Carlson, D.N., "Caligula's Floating Palaces," *Archaeology* 55, issue 3 (May/June 2002), 26.

36. On the Byzantine Yassi Ada ship which yielded the collection of tools, see Bass, G.F. and F.H. Doorninck, *Yassi Ada I. A Seventh Century Byzantine Shipwreck* (College Station: Texas A&M University Press, 1982); on the Serçe Liman ship and its cargo of Islamic glass, see Bass, G.F., B. Lledo, S. Matthews, and R. H. Brill, *Serçe Limani, Vol 2: The Glass of an Eleventh-Century Shipwreck* (College Station: Texas A&M University Press, 2009).

37. Wachsmann, "Deep Submergence," 130.

38. Accounts of the discovery: Bass, G.F. *Archaeology Under Water* (New York: Praeger, 1966) 134f.; see also Delgado, J. (ed.) *Encyclopedia of Underwater and Maritime Archaeology* (New Haven: Yale University Press 1998) s.v. Antikythera Wreck.

39. See his definitive essay on the mechanism in Price, D. de S., "Gears from the Greeks: The Antikythera Mechanism—A Calendar Computer from ca. 80 B.C.", *Transactions of the American Philosophical Society* (New Series) 64, no. 7 1974: 1-70; the claim about its importance is made in Price, D. de S., *Gears from the Greeks: The Antikythera Mechanism—A Calendar Computer from ca. 80 B.C.* (New York: Science History Publications 1975), 48.

40. Freeth, T., Jones, A., Steele, J.M., Bitsakis, Y. "Calendars with Olympiad display and eclipse prediction on the Antikythera Mechanism," *Nature* 454, (July 31, 2008): 614-617.

41. White, L. "Technological Development in the Transition from Antiquity to the Middle Ages," *Tecnologia, economia e società nel mondo romano: Atti del Convegno di Como, 27/28/29 (Settembre 1979.* Como, 1980): 235.

42. Cicero, *Tusculan Disputations*, 1.25.

43. Vitruvius, *on Architecture* 9.8.2-7.

44. Diodorus Siculus, *History* 14.42-50. Marsden 1971 gathers the surviving technical treatises on Greek and Roman artillery in two volumes. On the technology used in ancient Greek starting gates, see Valavanis, P., *Hysplex: The Starting Mechanism in Ancient Stadia: A Contribution to Ancient Greek Technology* (Berkeley: University of California Press, 1999).

45. Heron, *Pneumatics*, 1.38.

46. Aulus Gellius. *Attic Nights* 10.12.8-10.

47. Heron, *Pneumatics*, 1.15-16.

48. See Green, *Alexander*, 469.

49. Homer, *Iliad* 18.369-379.

50. Athenaeus, *Philosophers at Dinner* 5.198e-f.

51. Polybius 12.13.11.

52. *Civil Wars* 2.20.147.

53. James, P. and Thorpe, N. *Ancient Inventions* (New York: Ballantine Books, 1995), 138-140.

54. On Al-Jazari's treatise, his inventions and his legacy, see Hill, D.R. (trans.) Ibn al-Razzaz Al-Jazari, *The Book of Knowledge of Ingenious Mechanical Devices* (Holland: R. Reidel Publ. Co., 1974).

55. I am leaving aside for now consideration of the Baltic and other bodies of fresh water, which can also provide excellent conditions for preservation, but have yet to yield archaeological discoveries to rank alongside the Antikythera, Uluburun, or other famous ancient shipwrecks of the Mediterranean.

56. At the time of writing, Parker, A.J. *Ancient Shipwrecks of the Mediterranean and the Roman Provinces* (Oxford: Tempvs Reparatvm, 1992) provides the most comprehensive gazetteer and overview of ancient Mediterranean shipwrecks. On the recovery of other Hellenistic bronze statues from Mediterranean shipwreck contexts, see Hemingway, S. *The Horse and Jockey from Artemision: A Bronze Equestrian Monument of the Hellenistic Period* (Berkeley: University of California Press, 2004), 17-19.

57. Suetonius, *Life of Domitian*, 18.

58. Canfora, L. *The Vanished Library: A Wonder of the Ancient World* (Berkeley: University of California Press 1990) provides a good basic introduction to the history of the library and legends about its destruction.

59. Grant, M. *Cleopatra—A Biography* (Book Sales: Northfield 2004 (first published 1972)), 71; few have followed Grant's suggestion that only the book warehouses were burned, but the Alexandrian library clearly endured after Cleopatra VII and so cannot have been wholly destroyed. No report can be confirmed, but Aulus Gellius put the number of books in the Alexandrian library at 700,000 (*Attic Nights* 7.17) and says they were all burned accidentally in Caesar's Alexandrian war. Seneca, however, thought that only 40,000 books were burned (*On Tranquility of Mind* 9.5).

60. Galen 17, no. 1: 607.

61. Photographs in Bass 2005: 126.

62. Bass, *Beneath*, 47. The presence of small finds inside amphorae may be due to the actions of the humble octopus. This author has heard several stories from Institute of Nautical Archaeology divers who have witnessed the acquisitive and territorial activities of these cephalopods on Mediterranean shipwrecks.

63. Personal communication with the author, 2008. See also Wachsmann, "Deep Submergence" on deep submergence excavation technologies and techniques, and the importance of not letting lack of research funding be an excuse to hand over the entire cultural heritage of the deep seas to commercially motivated treasure hunters.

64. Plutarch, *Life of Sulla* 26.1.

65. Sider, D., *The Library of the Villa dei Papiri at Herculaneum* (Los Angeles: J. Paul Getty Museum, 2005).

# Chapter 5

# Removing the "Grand" from Grand Unified Theories: An Archaeological Case for Epistemological and Methodological Disunity

## William H. Krieger

The *new archaeology*, an archaeological program of the 1960s, was the product of an unlikely collaboration between Archaeology and the Philosophy of Science. Based in part on philosophical models of scientific explanation and confirmation, created by logical positivist Carl Hempel, archaeologists believed that Hempelian explanation and confirmation would provide archaeological science the materials needed to found itself on unshakable ground. Hempel's original (Deductive Nomological, or DN) model of explanation made clear his view that complete explanations were valid, deductive arguments of this form.

| | | |
|---|---|---|
| $C_1, C_2, \ldots C_k$ | Antecedent conditions | |
| | | Explanans |
| $L_1, L_2, \ldots L_r$ | General laws | |
| | | Logical Deduction |
| E | Description of the empirical phenomenon to be explained | Explanandum |

According to American new archaeologists, archaeological explanations would follow this format, which would necessitate the existence of archaeological *general law statements* and *initial condition statements* that together would provide the necessary and sufficient conditions to deductively conclude the explanandum statement—the archaeological data to be explained. If this plan had worked, archaeological science would have triumphed over the problem of underdeterminism[1] and archaeology would have been capable of fully explaining the archaeological record and predicting future finds and work.

For a variety of reasons, both archaeological and philosophical (reasons which are not the focus of this chapter), this movement failed to transform archaeology into a deductivist science, and this failure left archaeology in its current exciting, albeit fractured, state.[2] However, despite new archaeology's epistemological problems, many archaeologists continue to support one of the tenets of Hempel's program: scientific unification. For Hempel, this meant that all sciences (by definition) could be recast in Hempelian terms, as Hempelian DN (or IS or DS) explanations.[3] If all science could be reduced to logic, as Hempel intended, and if the goals and methods of science could be systematized to realize this goal, then all science, from Physics to History, would be shown to be essentially the same task happening at different levels of organization. Unification was closely connected to Hempel's belief in scientific reduction, where higher-level sciences would eventually reduce to/be replaced by lower-level sciences, with the ultimate goal of a single unified science: Physics.[4] For archaeologists, unification was a field goal, and their vision was a grand unified theory of archaeology: one that standardized methodology and practice; a field cast along Hempelian lines, resulting in the sort of repeatability and predictability archaeologists believed would be necessary for good scientific practice.

For archaeology at least, this chapter's author believes that the position that a unified theory is the best way to achieve local scientific goals and practices can only result in a loss of proper focus for archaeologists. This chapter will show that archaeology is a field that should view epistemic disunity as a strength. In fact, archaeologists do their best science by resisting strong versions of scientific unification. If this is true for archaeology, the author would also raise the question of whether our unconscious assumption that grand unified theories (GUTs) should continue to be a central goal of the greater scientific community (or communities). The answer (in short) will be that a GUT's utility lies in the questions it answers, not in its ability to give metaphysical comfort to scientists and the general public.

# The Goals of Archaeology

Despite depictions in the media of archaeology as a unified, wholly scientific enterprise that pushes our understanding of the past backwards through the millennia, archaeologists disagree strongly about the methods and goals (and at times, about the point) of archaeology. This debate stems from differing epistemological movements within the discipline of archaeology (pre-theoretical, traditional, scientific, and interpretive).[5] More than a matter of academic minutia, the archaeologist's (or archaeological community's) beliefs about the nature of archaeological enquiry strongly influences the tools an archaeologist will use and in the data that s/he will collect, as well as in the explanation resulting from those data.[6] In addition, the author will argue that contemporary archaeological models, such as those by Alison Wylie and Peter Kosso, may allow the field to unify (in a sense), with details to follow.

*The pre-theoretical model*, called antiquarianism by twenty-first century archaeologists, focused on the acquisition of the past, either for personal or cultural enrichment. Collectors' interests, whether to have the best collections of antiquities, or to help to show a current culture to be the outgrowth of ancient empires (and the more ancient the connection, the stronger the foundation) drove them to search for the best materials the ancient world had to offer. This could be accomplished by discovering an ancient city and bringing its treasures back for the glory of the empire, or by traveling the ancient world in search of holy routes, cities, etc., for the purposes of setting up pilgrimage sites in the ancient world. Methods followed theory, even in a context that was supposedly pre-theoretical. Collectors hired scores of laborers to tunnel or carve through sites in search of the best remains, with no interest in securing a find's cultural or temporal context. Although more technologically advanced, modern treasure hunters would arguably fit this category, as the market drives their motives. Although some groups do keep very good records of their pieces, they do so not for the records' intrinsic worth, but as a way to avoid litigation or as a means to secure a better price for the recovered objects.[7]

*The traditional model*, en vogue at the dawn of the twentieth century, is now also known as culture history. Culture history focused primarily on site excavation, with the goal of collecting information in order to link specific finds and patterns of activity to real dates. As the goal of this archaeological paradigm was to catalog and understand local changes in order to set that site's chronology, this epistemological and methodological pairing makes perfect archaeological and philosophical sense. Although this form of archaeology is not practiced by the majority of contemporary archaeologists, many of its methodological advances became the foundation for contemporary models (see below). Among the practices that have been embraced by all are those techniques, such as relative dating by analyzing changing architectural styles and by collecting and categorizing a region's ceramic corpus, that would help archaeologists to better understand the minutia surrounding a particular people.[8]

*Scientific archaeology*, labeled processual archaeology or new archaeology, represented a change in archaeological focus, based on a desire for archaeology to become more scientific. Culture history, already moving in the direction of systemization, produced artificially static pictures of the past. To extend the photographic metaphor a bit, imagine a person coming into a city and taking a photograph of what s/he saw. Would this photo represent life in the city? Would it capture radical changes in power, politics, demographics, urban planning, etc.? Would one photograph tell the story of the whole city, or just of the small portion captured by the expert's lens? The processualist, interested in understanding a dynamic (as opposed to a static) people, might be thought of as shooting a video, creating a more complicated, though presumably truer account of the past. This new epistemological direction, initially guided by logical positivism (and in the Americas, specifically by the work of Carl Hempel) would bring archaeology, a nascent science, into line with the established sciences, such as physics and chemistry, with the goal that archaeology would be able to gain the predictability and repeatability, that the established sciences were thought to represent.[9] The positivistic ideal would be gained by turning focus away from specifics in order to find general rules (or laws) of archaeology. Like the laws of physics, these archaeological laws would be the foundation upon which the new archaeological science would stand. Archaeologists attempting to put Hempel's vision into archaeological practice quickly found that the idea of logical positivism was hard to put into archaeological practice, and American processual archaeology has moved in a number of directions since its start in the mid-twentieth century. In addition, processual archaeology was incorporated differently in Europe than America, embracing a combination of inductive and deductive models for archaeological confirmation, instead of the rigidly deductive models that proved so problematic for the American Hempelians.[10] However, regardless of the variety of processualism being pursued, this author would argue that the *methods* of processual archaeology were unified, in step with the program set by the Hempelians in the United States of America.

On both sides of the pond, processualism's theoretical shift necessitated practical changes that would have serious effects on the way that localized archaeologies were practiced. For example, the goals of the new archaeology created massive changes in what archaeologists considered data. As noted above, pre-theoretical archaeological work had as its goal the recovery of artifacts, and as such, the archaeologist looked for pieces of monumental architecture, precious metals, and the like. Culture historical archaeological work focused on setting chronologies, so aids to this end, including changing ceramic and architectural corpuses, as well as other artifacts capable of providing data to set relative and absolute dates were listed in site reports as being crucial to excavations. New archaeologists, attempting to break free of the artificiality of strict chronologies, turned to a host of new specialists and data sources in order to discover general laws of archaeology. These data sources included the aforementioned dating aids, as well as the talents of a wide variety of experts on soils, metals, bones, languages, and other materials newly the provenience of archaeologists. As processual archaeologists needed to gather large amounts of data (seen as the only way to create general laws), changes would need to be made on the ground in order to provide those data. Further, in order to make the data more representative of archaeology's new goals, general laws and higher-level explanations, what constituted good archaeological practice would need to change to reflect the new philosophical paradigm. As individual sites represent the idiosyncratic activities of singular peoples (or of individuals), the culture historians' focus on the excavation of individual sites could be seen as counterproductive to the construction of good science. As an emphasis on and excavation of sites (the small picture) would distort the new (more realistic) regional picture, sites were de-emphasized, becoming a part of the landscape and being excavated in that context. This is not to say that landscape studies and regional surveys are new to archaeology. Archaeologists have long given attention to landscape, especially within settlement archaeology. In recent years, however, the focus on landscape has shifted and what was once generally passive background has now assumed the foreground. This results partly from archaeologists expanding their view beyond individual sites to considering a more comprehensive distribution of human traces outside of places seen as loci of human activity, such as cities or cultic sites.[11] Although we will return to this subject presently, the idea that archaeology is a science, and that the search for universality and generality would lead to the repeatability, predictability, and certainty that are seen as hallmarks of good science, has resulted in a modified (closer to the European model) of processualism being a major player in today's theoretical landscape.[12]

*Interpretive archaeology*, sometimes called post-processualism by its proponents, is a term used to capture a variety of responses to processual archaeology. Although these theoretical models vary widely, they have in common a rejection of one of the major tenets of processual archaeology: that archaeologists are able to use a set of universal laws (or to some, any scientific apparatus) in order to explain a singular (true) past. Initially, this led some post-processual archaeologists to deny that archaeology is scientific at all. These anti-science archaeologists saw archaeology as a political tool, one which could be wielded by anyone desiring to use archaeological data to create a history and establish a connection to past empires and glory (self-consciously similar to the antiquarian ideal, discussed above). However, what began as a reactive, and largely negative, reaction to processual archaeology has grown into a variety of progressive programs, asking a variety of fascinating epistemological questions about the preconceptions (be they gendered, political, or other) built into our current models of science. As an interesting note, the theoretical questions asked by interpretive archaeologists have not translated into methodological changes on the ground. If archaeology is not a science, if general laws are no longer the goal of archaeology, and if the focus of archaeology thereby shifts back to the individual (be it a singular ethnic perspective or a particular piece of geography), then one would imagine that archaeological practices would change in some way to fit this model. However, even the most radical anti-scientific archaeologist still digs like a processualist, employing tight controls on data acquisition, employing a variety of experts from a variety of fields, including physics, chemistry, biology, and geology, and publishing their results in peer reviewed journals. Whether there will be an interpretive methodology in the future is an interesting question, but for now, data are still king, regardless of their eventual use (or misuse).

Finally, the question arises of whether a more nuanced position is possible (or preferable) to those mentioned above. This "new science" archaeology[13] would have at its epistemological heart a more contemporary understanding of the workings of science than its forebears. Certainly, science has significant problems, including biases based on a host of yet unseen underpinnings. However, to claim that the best answer to these issues is to wring our hands of science (or perhaps its neck) and fall to total subjectivity is absurd. This alternative, called by some a *multi-vocal* approach,[14] recognizes that a linear, step-by-step march toward certainty, based on untainted objective facts, is a fairy tale. However, although multi-vocality was argued by Trigger[15] as a way to avoid both the false objectivity of processualism and the total subjectivity of post processualism, contemporary scholars have used it to argue for an anti-science approach to archaeology. Authors (e.g., Trigger) that push for a multi-vocal approach argue that archaeology has failed to find an objective foundation. However, their conclusion, that archaeology is a failed science, misses Trigger's further arguments, showing that science (in general) never was able to live up to the strict objectivist standard, and that, like other sciences, failures to meet an unrealistic standard should not stand in archaeology's way as a scientific discipline. This author's alternative is a "new science" approach, where different perspectives (any that can be captured by data, data wielded by a variety of sources for alternative purposes) strive to tell their stories, and where a true (or truer) account is the result.[16] This last point, that something true comes from a series of potentially contradictory accounts, is what separates the *new science* from *multi-vocal* approaches. Where the latter would claim that each account be advanced by its own proponents, with the mob roar taking the place of a reasoned discussion, the former would claim that an integrative strategy—playing different accounts against each other, understanding them in context, looking for other sources of justification and weight, etc.—can lead to the sort of true account sought (but unattainable) from within the processual stream of archaeology. This approach is already regularly in use in History (ironically, a field that Hempel used as his test case in creating new sciences). Where scientists generally view the existence of a contradictory explanation as proof of the failure of the account under study, modern historians understand the value of a less rigid (and less deductivist) approach in capturing the story in any historical event. Readers learning about American history have no problem understanding that the historical event called the "discovery of America" is not captured in the Colonial account, nor in the English account, nor in the Indigenous American account. These three (plus many as yet unknown versions) totally contradictory histories, put together, provide students with a more accurate understanding of the 1400s than would any unified account without demanding an "anything goes" subjectivity as the price of rejecting the traditional story.

Archaeology, depending both on a variety of potentially tainted sciences and on other equally problematic data sources (such as written and oral histories, ethnographic analogy, political, gender-based, religious, and ethnic agendas, etc.), could not be better placed to explore this *new science* approach than perhaps any other science. Whether this epistemological platform demands a change in methodology will be a question answered below.

# Theory, Methods, and New-Science

The preceding sections have concentrated on the effect epistemological ideals have had on scientific practice, and on the need for further change in the relationship between archaeological theory and methods. Archaeologists, like all scientists, have seen rather large-scale changes occur in their lab and fieldwork, based on changes to the reigning theoretical model. One example of this trend in archaeology is in the relative percentage of archaeological work that was done as site excavation, as opposed to regional survey. Alluded to above, theoretical changes leading to the rise of processual (and later responses to processual) archaeology created a demand for a range of new data sources. Archaeologists interested in creating general laws of archaeology, followed more established sciences in searching for laws. Hempel believed that laws were necessary for proper scientific explanations, and that confirmation (choosing between potential explanations) could not occur without a robust scientific structure, built upon that nomothetic foundation.

Although Hempelian science was the primary focus of American processual archaeologists, his work was by no means the only material being read by scientifically minded archaeologists in the middle-twentieth century. For example, Karl Popper who was a contemporary of Hempel's, repudiated Hempel's beliefs about the possibility of incremental confirmation, although both depended on the existence of universal laws used to test whether a particular explanation candidate was properly scientific (to Popper, "falsifiable"). Others who did not agree with Hempel's posit—that science is necessarily unified and universally true—still demanded the accumulation of data on an as yet unprecedented scale. Some, such as Thomas Kuhn, posited that fields do not become sciences until they are unified behind a single, unconsciously agreed upon worldview. Before this can occur, factions fight amongst themselves for supremacy, each trying to find a set of universal principles.[17] Imre Lakatos disagreed with Kuhn on the existence of shared worldviews, instead believing that different factions present worldviews (each made up of a central core and a changeable surface) that constantly battle each other for supremacy. As these sorts of positions do not posit the existence of a central core of truth, there can be no universal standard of right or wrong.[18]

A separate question that is rarely a topic of study (and that remains unanswered) is whether Hempel's model of good science (one that begins with a theoretical ideal and then seeks to apply it to scientific practice) is the way that science should work. Whether or not this *should* be the case, the new (scientific) archaeologists claim this *is* the way they work. They argue that the theoretical changes that guide processualism (including a push for more data, and for more types and sources of data), must result in particular methodological changes to the way archaeology is practiced worldwide.

In fact, I have argued elsewhere (Krieger 2006) that regional geography shapes the theoretical landscape as much as epistemological theorizing shapes the practice of archaeology. Further, the changes demanded by the new archaeologists were initially driven by local needs, and not general theoretical demands at all.[19] The question is: how did local decisions become re-cast as theoretically driven methods? Briefly, American theorists wished to maximize the data that they were collecting with the hope that they would discover general (universal) laws of archaeology, similar to those thought to be abundant in the natural sciences; they shifted their focus from excavation to survey. This was done because American archaeologists rightly analyzed their regional geography, and recognized that sites (generally lightly stratified) took an enormous amount of time and resources to excavate, and resulted in little data, while regional survey—which was far less costly in terms of time and resources—resulted in much more data. Other reasons for the switch to survey were practical: focusing on the limited amount of both time and resources available for the exploration of large areas, and on shifting budgetary priorities for the government and other sponsors of archaeological work. Yet others were theoretical, focusing on new ideas about the nature of archaeological preservation. For instance, in regions with highly stratified societies, important sites would be under the control of the wealthy or powerful, while the majority of the populace would live in surrounding areas. As such, site-centered archaeology would produce a distorted, wealth-centered picture of a period. In addition, areas such as the American Southwest had a highly heterogeneous population. So, concentrating on any one area would provide information on one small group to the expense of the many others in the surrounding areas. The lack of natural boundaries also made it difficult for site-centered archaeology in the Americas. Without natural boundaries, it was difficult to determine where a site ended. Archaeologists were forced to set artificial boundaries between sites (stopping work when they were unable to find a subjective number of artifacts per square meter). As a result of these concerns, American archaeologists decided that survey would allow them to focus on recognizing large trends, or general laws, as opposed to the site-centered approach of the past, which could only provide very specific laws relating to small groups (or portions of groups).

As the theoretical and methodological goals of the new archaeology took hold, first in the Americas, and then throughout other regions where American scholars had influence, the theoretical outlook was packaged along with the techniques that were proving effective in the Americas. For instance, some of the largest and best known modern field schools in Israel (Gezer and Miqne/Ekron) were co-run by American archaeologists, and there was talk on these digs about the need for Israeli archaeology to catch up to the theoretical models of the Americans, and this was also reflected in senior scholars referring to themselves at conferences and in publications as new archaeologists.[20]

Tellingly, those archaeological theoreticians who identified processualism with survey claimed that their model was an example of theory informing methodology, which is in line with the ideals of logical positivism. In fact, the regional geography of the Americas informed the processualists' methods as much as did their theoretical outlook. Had they been aware of the connection between local needs and theoretical decisions, they would have realized that places with differing regional geographies would need to employ methods with a different focus in order to maximize their data. Of course, detractors could argue that archaeology should be unified in its practices. This argument would be in line with the idea that archaeology is setting out to find universal laws; however, in order to make this claim, they would need to argue that the landscape has no bearing on the reasons that people would settle in an area. Given the radical differences between settlement patterns in the Americas and the Ancient Near East— patterns that are easily explainable by referencing differences in regional geography—that argument seems untenable.[21]

In fact, the regional geography of Israel could not be more different than it is in the Americas (despite the fact that the climate in Israel is very similar to that of Southern California). Unlike the sites in the Americas, sites in the Ancient Canaan are much easier to define, as the landscape and political facts on the ground forced people to live in more structured and more compactly located areas. These sites, called *"tels" (or "tells")*, are highly stratified and contain, at times, evidence of 3,000 or more years of continual occupation. Although some groups did not live at tells, a large enough percentage lived there to make site-centered archaeology far more important than it was in the Americas. In addition, as the groups in this region were far more homogenous than in the Americas, understanding key sites in the region would be an effective way to understand the peoples living in them. As a result of these regional differences, the archaeologists working in Israel found themselves unable to make the methodological shift to survey that the American school demanded. Theoreticians working in Israel saw a massive increase in the amount of data coming in, due to the addition of many specialists on every archaeological project. However, they spoke of themselves as being backward or behind the times, due to their (mistaken) perception that their way of doing archaeology was antiquated and unscientific—not understanding that their conscious decision to reject the straight survey approach, which would have been as destructive in Israel as a straight excavation approach would be in the Americas, was as theoretically sophisticated as was the Americans' decision to go for that approach.[22] This issue becomes even more striking when archaeologists look to even more disparate ways of "doing" archaeology. Deep-sea archaeology (See Buxton, this volume) is a field that has little in common with the theoretical or methodological mainstream in archaeology. However, the data generated by these archaeologists is as unquestionably important as it is archaeological. Underwater archaeologists, such as Jacob Sharvit of the Israel Antiquities Authority, have also staked out their own theoretical and methodological territory, on the one hand surveying and excavating shipwrecks (which are similar to American single stratum sites), while on the other dealing with stratified materials being inundated and disturbed by shifting sands and currents.[23] Given the clear role that geography plays on methods, the idea that these archaeologists should consider their methodology to be inferior to the American "ideal" is clearly absurd.

## From Geography to Epistemology

Rather than blaming archaeologists for an intellectual lapse, however, the confusion here points out a systemic failing in logical positivism—the model championed by American processualists (in archaeology and in a number of other social science fields that were recasting themselves as sciences in the early to mid-twentieth century) as the key to scientific objectivity. In fact, even in Hempel's day, other philosophers of science[24] had already come to the conclusion that science does not have a particular method. Despite the fact that every basic science text has a section on "the scientific method," Hempel's assertion—that scientific theory can proceed toward certainty, or that there is a particular unified direction to science—has been contested by many philosophers of science, ranging from those who believe that there is no method at all to science (Paul Feyerabend and Imre Lakatos) to those who believe that methods are matters of particular choice, and are incorrect, albeit useful (Thomas Kuhn, Michael Scriven, and Nancy Cartwright). Although each of these theorists believed Hempel wrong for different reasons, and each sought to move science in different directions, using examples from different sciences, they nonetheless understood this problem as significant.

Archaeologists also continued to struggle with archaeology's attempt to step into the scientific mainstream. In the 1980s, two polar positions, based on processual (now named neo-processual)[25] and post-processual (renamed interpretive) archaeologies,[26] sought to correct original shortcomings and to end the debate as to whether archaeology was or was not a science. Although many in the archaeological community continue to this day to fight either for or against scientific archaeology, a number of theorists, from both philosophical (including Peter Kosso and Alison Wylie) and archaeological (such as Bruce Trigger) camps have put pen to paper asking whether there is a real difference between the two accounts (once invective and straw man positions are removed from serious debate). Kosso shows that there is a way beyond choosing one or the other static point of reference. Kosso is clear in his position that archaeological objectivity is possible: "Some of what we claim to know about the past is true; the rest is false. The purpose of this book is to describe ways of telling the difference."[27] At the same time, he clearly believes that there is a great deal of archaeological interpretation going on under the guise of fact. Given the fact that both the science and interpretive camps are well aware of these issues, Kosso asks whether there is a real difference between Binford and Hodder's approaches, showing that it is possible to use their own words to create a consilience position that better represents archaeology as it is practiced on the ground.[28] Wylie uses a tacking metaphor to show how local and general issues both have a place in archaeological discussions (in many cases, in the same archaeological discussions), and she shows throughout her work that allowing in more information does not necessarily lead to a scientific breakdown: ". . .the key to understanding how archaeological evidence can (sometimes) function as a semiautonomous constraint on claims about the cultural past is to recognize that archaeologists exploit an enormous diversity of evidence—not just different kinds of *archaeological* evidence, but evidence that depends on background knowledge derived from a number of different sources, that enters interpretation at different points, and that can be mutually constraining when it converges, or fails to converge, on a coherent account of a particular past context."[29] Finally, Wylie's work on standpoint epistemology, applied to archaeology, shows the possibility of a robust "unified" way forward.

# Conclusion: E Pluribus Unum

There are many positives to unification on a psychological, and even on a scientific, level. Having a grand unified theory calms us. It makes us feel like we are in command of a universe that otherwise is clearly well beyond our comprehension, much less control. Having a unified method to science, even if that method is wrong, at least gives us an agreed upon starting point for scientific inquiry, what Cartwright would call "organizing principles."[30] And, in some cases, there does seem to be evidence that a certain result can be best explained by subsuming it under a larger law. I would posit that in these cases, when unification advances knowledge, unification is a good thing. However, methodological unification for unification's sake can be detrimental to scientific progress.

In the case of archaeology, as shown above, the processualists attempted to reform archaeology under a unified set of methods. Although this goal made some sense (problems with its implementation aside), success would have meant a full stop to archaeology in Israel, and probably to archaeology in other areas of the world. Instead of moving to a unified set of methods, Israeli archaeologists have unwittingly given some proof to the aforementioned theoretical writings of Wylie and Kosso. In Israel, we see real (if messy) progress from anti-unification trends, based on regional geography. In this case, the use of site excavation, considered anachronistic in the Americas, delivers local knowledge that would have been obliterated by a survey-only approach.

The final question is whether this rejection of universal laws in archaeology forces us to abandon any concept of objectivity—as some in the interpretive camp have claimed. If the Hempelian ideal is not the point of archaeology, it is difficult to resist the claim that archaeology should only focus on the local, and that this field is not really a science at all. However, a good case can be made for a robust archaeological science, one that is capable of handling the controversies and contradictions found in the archaeological record.

The *new science* model, taking its cues from Trigger's call in 1984 for a more nuanced approach, and seeing steps toward that goal in Wylie's position on standpoint and Kosso's on epistemic independence, can have its epistemological cake and eat it, too. Unlike the polar positions, the new science approach does not set unrealistic (in practice or in principle) epistemic goals. Instead, the *new science* approach does exactly what processualism (in its earliest ideations) wanted to achieve[31]: it provides a vivid understanding of the past in the place of the static picture provided by culture history. It does this, without falling into the artificial formality of deductive science or into the subjectivity of anti-processual historicity. While their solutions, standpoint and epistemic independence, are able to avoid these issues as well, I would argue that both positions have problems strong and relevant enough to countenance the creation of another position.

Standpoint theory is the idea that scientific unity (understood to reside in the masculine view of the world), must give way to a discussion between stakeholders representing multiple positions in (and out of) a scientific community. Although this seems to deliver on this chapter's goals, some standpoint authors have used their position to argue that certain groups have epistemic privilege over others.[32] This is problematic for two reasons: First, it replaces the epistemological superiority of one group with another. At best, privileging one nondominant standpoint ignores other oppressed groups, ignores positive elements in the dominant position, and ignores the fact that every group is blind to its own biases.[33] The oppressed are (at times) capable of seeing problems in the majority's position (problems that are overlooked by those in power) but they are just as blind to their own biases, weaknesses, and hypocritical positions as are those they hope to replace. Second, standpoint implies that there is something static about a person's position in the community (as a woman, as an ethnic minority, as a person of low socioeconomic status, etc.), and that this one characteristic is what truly (and in all cases) defines that person. In fact, a person easily represents a multitude of (at times contradictory) standpoints, each of which becoming partially dominant or dominated several times during the day. The layers of complexity that real life adds to standpoint theory would need to be explained (from within this model) for it to be able to serve as a true foundation to scientific inquiry.

Kosso believes that the failed monolith that is logical positivism should be replaced by epistemic independence—the idea that objectivity can (in a sense) be recaptured by looking toward a variety of sources of data, each giving its own reason to believe the conclusion. This idea, that what can be missed by a myopic focus from one perspective can be captured by a convergence of data, sounds right, as did standpoint at first blush. However, this concept suffers from questions as well. First, there is a real worry about whether epistemic independence ever really exists (a point that Kosso speaks to at some length, and his conclusion that epistemic indepencence should be seen as more a continuum than a strict standard is absolutely on point). Equally problematic, this author worries that epistemic independence relies on a sense of objectivity that is a bit too robust, depending on methodological unification to provide the structure for epistemic independence. If we can be so confident of our scientific apparatus, then many of the worries that have been the subject of this chapter would not exist. Kosso seems to depend on just enough methodological unification to make even the weaker sense (discussed above) of epistemic indepdendence virtually impossible to achieve.

Having made these critiques in brief, critics of my position could argue that my position is a result of mere hairs being split unnecessarily. Although this author would argue that there are substantive differences between standpoint, epistemic independence, and new science, he would be most happy for those differences to be addressed, and for an unified (so to speak) approach to this issue to be the result. Whether that resulting approach will be a variant of Wylie's or Kosso's or a new-new-science, the picture that this author sees when he looks to the future of archaeology is promising and bright. Once the bumps in the road can be smoothed over, this new approach (one that seems to be converging, given the outstanding work by Wylie and Kosso) will be able to answer the charges of radical deconstruction, and rest on an epistemic foundation that is both true enough and strong enough to support itself. The result, which is slowly coming into focus, is a more accurate (if more complicated) account of human history.

# Notes

The birthplace of this chapter (as well as the inspiration for this volume) was the NEH Summer Institute on Science and Values in 2003, Led By Professors Machamer and Mitchell of The University of Pittsburgh. A draft version of this chapter was presented at the International Three Societies Meeting (joint meetings of the British Society for the History of Science, the Canadian Society for the History and Philosophy of Science, and the History of Science Society) in Oxford, 2008. The author would like to acknowledge the sponsoring organizations, and to thank the organizers and the participants for their feedback. In addition, the author would like to thank the following groups for allowing access to their data or for funding, both of which were crucial to the completion of this chapter: Tell el-Farah, South Excavations, the Israel Coast Exploration Project, and a number of groups and departments at the University of Rhode Island, the Archaeology Group, the Hazen White Center, the Philosophy Department, the College of Arts and Sciences, the Provost's Office, the Center for the Humanities, and the Foundation. Finally, the author sends love and thanks to my wife Aliza and my daughter Ella for their unending supply of support and distractions. (As to who provided which, I will not comment.)

1. This problem, that the set of explanans sentences in an Hempelian explanation can never provide sufficient means to deductively conclude the explanandum sentence, is called equifinality in archaeological circles.

2. Among its chief problems, the new archaeology was unable to meet Hempel's strict standards for deductive explanation and confirmation. Although this was seen as a particular failing on archaeology's part (a result of archaeological underdeterminism), in fact, this was a problem for all Hempelian science. For a detailed look at these issues in their historical context (from an archaeological and philosophical perspective) see Krieger, *Can There Be a Philosophy of Archaeology: Processual Archaeology and the Philosophy of Science?* (Lanham: Lexington Books, 2006).

3. Although Hempel is most famous for his DN model, Hempel later introduced two other types of explanation, Inductive Statistical (IS) and Deductive Statistical (DS), to fill out gaps in his original model.

4. For Hempel's work on reduction (and to show the role that unification and reduction play in his philosophy of science) see Hempel, *The Philosophy of Natural Science* (Upper Saddle River: Prentice Hall, 1966).

5. For more information on the shift from pre-theoretical archaeology (sometimes called antiquarianism) to the beginnings of the modern discipline, see Schnapp, *The Discovery of the Past* (New York: Harry N. Abrams, 1997). For a good (and unbiased) treatment of the different strains of archaeology, see Trigger, *A History of Archaeological Thought* (Cambridge: Cambridge University Press, 1989). To get an understanding of the epistemological issues behind these movements, see Krieger, *Philosophy of Archaeology?* (Lanham: Lexington Books, 2006).

6. Although these movements are presented chronologically, each model has its contemporary proponents.

7. There are interesting arguments to be made on the subject of treasure hunting in modern archaeological contexts, but these go beyond the scope of this chapter.

8. As a side note, although this approach is not practiced by many in the field, it is this model, as well as the pre-theoretical model, that the public thinks of when it hears of archaeology.

9. Hempel's model of explanation (for the traditional sciences, and he thought, for history and other new sciences) was to be deductive, following the DN pattern mentioned above. Hempel expanded his thoughts on explanation to include a Deductive Statistical

model and an Inductive Statistical model as well, but all of his models require a strict connection between initial conditions and universal laws.

10 . To read about the British version of the "new archaeology" from one of its pioneers, see Clarke, David *Analytical Archaeology* (London, Methuen Press, 1968). For an excellent account of their impact, see Bell, James. *Reconstructing prehistory: Scientific method in archaeology* (Philadelphia: Temple University Press, 1994).

11. For more on these changes, see Ashmore and Knapp, eds., *Archaeologies of Landscape: Contemporary Perspectives* (Malden: Wiley-Blackwell, 1999).

12. This despite the problems archaeologists would have understanding the relationship between theory and practice, resulting in the problems addressed by this chapter.

13. The author recognizes that this title is totally inadequate and hopes for suggestions to better it. However, the alternative, using a much nicer title already in use (and already loaded down with its own baggage), was unacceptable.

14. For a good account of a contemporary archaeologist who sees multi-vocality (in the contemporary sense) as central to his program, see Hodder, *The Archaeological Process: An Introduction* (Oxford: Blackwell Publishers, 1999).

15 Trigger, "Alternative Archaeologies: Nationalist, Colonialist, Imperialist," *Man* 19 (1984): 355-370.

16. Both processual and interpretive archaeologies (in their more nuanced forms) claim the ability to handle multiple accounts of the past. Kosso rightly points this out, and shows that the real differences between processual and post-processual accounts seem to be disappearing. However, this author would argue that both positions have been painted (perhaps unfairly) so rigidly, both by detractors and by the proponants of each side, that a novel idea might serve better than a new layer of paint over an old façade.

17. See Kuhn, *The Structure of Scientific Revolutions* (Chicago: the University of Chicago Press, 1970) for his argument against the positivistic conception of science, and for his explanation of the evolution of science.

18. As there is no way (a priori) to judge between competing theories, questions about theory choice (and scientific progress in general) have to reference more complicated stories.

19. Ironically, the material being used as evidence for the promise of the New Archaeology in fact shows the need for, and existence of, methodological disunity.

20. A great example of this can be found in the foundational work of Archaeologist William Dever. A pioneer in the implementation of the new archaeology to Israel in the 1960s and 1970s (see Bunimovitz and Greenberg, "Of Pots and Paradigms: Interpreting the Intermediate Bronze Age in Israel/Palestine." Pp. 29 in *Confronting the Past: Archaeological and Historical Essays on Ancient Israel in Honor of William G. Dever*, edited by Seymour Gitin (Winona Lake: Eisenbrauns, 2006)). Prof. Dever became more critical of this program over time (Gitin and Dever, eds., *AASOR 49: Recent Excavations in Israel: Studies in Iron Age Archaeology* (Winona Lake: Eisenbrauns, 1989), 150), and has more recently (Dever, William. *What Did the Biblical Writers Know & When Did They Know It?* (Grand Rapids: William B. Eerdmans Publishing Company, 2001), 62-66) embraced Hodder's post-processualism.

21. See Rainey, "Historical Geography: The Link between Historical and Archeological Interpretation." *The Biblical Archaeologist* 45, no. 4 (1982): 217-223, for an excellent introduction to the convergence of these fields.

22. Of course, it is worth noting that both surveys and site excavations are practiced in both the Americas and modern Israel. In the Americas, sites are excavated to answer particular questions, such as those surrounding a locality's power structure or centralized food supply. In Israel, a number of surveys have been carried out for a variety of reasons,

including finding ancient water and agricultural systems, or understanding trade routes or ways of life for nomadic peoples.

23. Sharvit has developed his own method of dealing with these issues by inventing a square, caisson style, box that can be placed at an underwater stratigraphic feature. Once the interior is excavated, slightly smaller boxes can be placed inside to allow for deeper excavation. This system is being used in Caesarea and Akko.

24. For a radical example, see Feyerabend, *Against Method: Outline of an Anarchistic Theory of Knowledge* (London: Humanities Press, 1975).

25. See Lewis Binford's groundbreaking work in the American Southwest. In his theoretical work, he makes the case for archaeology to move beyond the subjectivity packed into telling local stories, and in a number of case studies, Binford shows exactly how archaeologists can use the new archaeology to create (regional) general laws and test their hypotheses. For an interesting example, a debate on the function of a regionally bound series of pits, see Binford, "Smudge Pits and Hide Smoking: The Use of Analogy in Archaeological Reasoning." *American Antiquity* 32, no. 1 (1967): 1-12; Munson, "Comments on Binford's 'Smudge Pits and Hide Smoking: The Use of Analog in Archaeological Reasoning.'" *American Antiquity* 34, no. 1 (1969): 83-85; and Binford, "Archaeological Reasoning and Smudge Pits—Revisited." Pp. 52-58 in *An Archaeological Perspective*, edited by Lewis Binford (New York: Seminar Press, 1972).

26. Many examples of these approaches exist. An excellent example of a range of multi-vocal approaches can be found in Habu, Fawcett, and Matsunaga, eds. *Evaluating Multiple Narratives* (New York: Springer, 2008), 29-44.

27. Kosso, *Knowing the Past: Philosophical Issues of History and Archaeology.* (Amherst: Humanity Books, 2001), 11.

28. Kosso, *Knowing the Past*, 71-72, is not arguing that consilience must result from his position, only that it is possible to see Binford's middle range theory as similar to Hodder's hermeneutic approach, despite the different goals of these enterprises.

29. Wylie, *Thinking From Things*, 192

30. This argument is the subject of Cartwright, *How the Laws of Physics Lie* (New York: Oxford University Press, 1983).

31. For some of the earliest, clearest statements about this goal, see Wissler, "The New Archaeology." *Natural History* 17 no. 2 (1917): 100-101, and Taylor, "A Study of Archeology." *American Anthropologist: American Anthropological Association* 50, no. 3, (1948 part 2).

32. See Harding, "Rethinking Standpoint Epistemology: What is 'Strong Objectivity'?" Pp. 49-82 in *Feminist Epistemologies*, edited by Linda Alcoff and Elizabeth Potter (NY and London: Routledge, 1993). Although Standpoint has changed quite a bit since this essay, many contemporary critiques still reference this landmark article, and newer authors, e.g., Saul, *Feminism: Issues and Arguments*. (Oxford: Oxford University Press, 2003) continue to focus on women having an epistemic privilege as central to standpoint theory.

33. Historically, oppressed groups have called for their voices to be heard, and in many cases, when those groups come to power, the first thing that they do is oppress the groups that once dominated them. Examples of the oppressed becoming the oppressor range from conflicts between ancient Greek city-states to the rise of Christianity in the dawn of the Holy Roman Empire, to modern conflicts in Bangladesh, Rwanda, and Iraq.

# Chapter 6

# The Virus As Metaphor In American Popular Culture, 1967-2010

## Steven C. Hatch

Fifty or more years from now, when the students of medical history look back at the major infectious scares of the late twentieth and early twenty-first centuries, they will inevitably wonder what all the fuss was about in 2009. After all, there was no previously unknown virus bursting onto the scene (like, say, SARS in 2003), killing people indiscriminately in its path, leveling entire communities. Indeed, influenza, the annual pestilential virus, was quite mild, killing only just over 2,000 in the United States over a twelve-month span, less than a tenth of the annual influenza mortality.[1] And yet any textbook written that far into the future will have to include a discussion about the 2009 "novel H1N1" influenza pandemic, either to discuss the public health measures that took place, the virulence of the virus (or lack thereof), or the impact of globalization on its spread. Regardless of what form the discussion takes, the 2009 influenza pandemic will likely go down as the great plague that wasn't, the virological bullet that humanity dodged through pure luck—and very much unlike the H1N1 influenza virus that had emerged nearly one century before, in 1918, when tens of millions of people worldwide were struck down in a matter of months.

## The Not-So-Great Influenza

Historians will also note that, lethal or not, the 2009 influenza pandemic created a good deal of hysteria, not least because a good many public health experts and virologists were concerned that this virus would all but re-enact the events of 1918. Yet popular reaction to the 1918 pandemic (also called "The Great Influenza"), while generating a good deal of frenzy in its own time, was structurally and thematically different from that of the 2009 pandemic, owing to instantaneous forms of mass communication, a more thorough public acquaintance with exotic viruses, and a different level of faith in science's ability to solve health crises. If one is to argue that the public reaction to the 2009 influenza pandemic belongs to a particular era, one is obligated to pick an arbitrary point in time, and that pick is subject to scrutiny in a variety of ways: go too far back in time (e.g., to 1918) and comparisons become meaningless, the analysis morphing into an ahistoric observation that contagion produces fear; come too close to the present and trends can be missed due to myopia.

That said, one arbitrary date that might serve as a good starting point to understand the reaction to the 2009 pandemic—what we might think of as "The Not-So-Great Influenza"—is 1967. By that year, two factors put in motion a set of anxieties that have found expression in American popular culture to this day, and reflect the peculiar anxieties of Americans during this period. The fact that as relatively harmless a virus as the novel H1N1 strain of 2009 created such a hubbub speaks more to those anxieties than it does to the virus itself. Thus, the virus becomes more-than-virus, a metaphor to illustrate the preoccupations of the citizenry.[2]

# Television, Viruses, and Metaphors

One of those factors was the rise of mass media in the form of television. While one could choose virtually any year from the 1960s to illustrate this, 1967 marked the point where the first of the modern, lethal viruses, the Marburg virus, was discovered. (Marburg and Ebola are the two viruses that comprise the *filovirus* family.) Marburg appeared on the scene as truly worldwide mass communication was coming into its own, and although it would receive a relatively small amount of attention at the time (it garnered a total of 236 words in two brief dispatches in *The New York Times*),[3] the outbreak would highlight the metaphors for which virus outbreaks would be used again and again in the coming decades: an unease with the Promethean incursions into nature in the name of science, and nativist misgivings about immigration, especially from places where peoples' faces are, in general, not white.

Some background: Marburg first came to attention in the west after an outbreak of hemorrhagic fever among laboratory workers in that eponymous German city.[4] The infected staff had been working with African grivets from Uganda; 25 workers were directly infected, and an additional six people were "secondarily" infected mainly from accidental skin pricks from needles while drawing blood.[5] Overall, seven people died. While this would prove to be a minor outbreak with a relatively low mortality rate,[6] it would become the archetypal viral outbreak, where nearly all the features required for mass panic would be in place: a virus comes from the jungles of Africa, taken to "civilization" by scientists doing biomedical research, yet some procedural snafu sends the situation horribly awry, so that not only the scientists but even innocent people are mowed down by the faceless pestilence.

Indeed, the only really substantive change in the storyline for the multimillion dollar-grossing movie *Outbreak* (1995)[7] is how the virus spreads: the fictional "Motaba" virus passes person-to-person by inhalation, not through contact with blood, allowing the transmission to be amplified to alarming heights virtually instantaneously. Motaba makes its way to the United States via the black market, where an African monkey is shopped around to pet stores by an animal quarantine employee. Not long after the monkey and his smuggler leave quarantine, people start turning up dead of a mysterious illness in the sleepy rural California town of Cedar Creek, and the outbreak is on.

Though fictional, *Outbreak* is loosely based on actual events described in Richard Preston's bestseller *The Hot Zone* (1994).[8] Similar to the movie, virus-harboring monkeys are shipped to quarantine in the Washington DC suburb of Reston, Virginia. Monkeys slowly begin dying of a strange infection, and samples sent to the CDC are subsequently identified as the highly lethal (though only transmissible through blood) Ebola virus that had previously only been encountered in rural Africa. By the time the alarm is sounded, however, there is ample evidence that this strain of Ebola is airborne. Preston chronicles the events leading up to and following the discovery of the virus that would eventually be named Ebola Reston, which included the killing of over 500 animals in the facility (four people became infected with the virus but would not become ill, the Reston strain being apparently adapted to killing monkeys but not humans).[9]

# The Morals of the Stories

On the surface, *Outbreak* and *The Hot Zone* both carry dire warnings about humanity's culpability in the emergence of these viruses due to ecological devastation (implicit in the former, underlined in the latter). To utilize the language of pop virology, the logic of the book and the movie is that if one plays with something "hot," one will get "burned." Since "one" is a technological, industrial culture, most everyone alive is part of "one." In the logic of these stories, a rainforest is the "hottest" place on earth, a virtual Pandora's Box of exotic, lethal microbes. We challenge these areas by encroaching on them (building highways, urbanizing, and whatnot) at our peril. In *Outbreak*, one of the scientists investigating the Motaba virus notices a medicine man on a hillside burning leaves in what appears to be a religious ritual. The scientist's African colleague explains that the man is trying to appease the Gods of the rainforest, who produced the destruction. "The Gods were angry," says the colleague, referring to the highway project that put the villagers in contact with the darkest (and thus deadly) parts of the jungle. "This was a punishment."

The anthropomorphization of Nature, merely hinted at in *Outbreak*, is turned into a moral in Preston's nonfictional account. *The Hot Zone* concludes its story by predicting that Ebola or HIV is only a prelude to something far worse:

> I suspect that AIDS might not be Nature's preeminent display of power. Whether the human race can actually maintain a population of five billion or more without a crash with a hot virus remains an open question. Unanswered. The answer lies hidden in the labyrinth of tropical ecosystems. AIDS is the revenge of the rain forest. It is only the first act of revenge.

While the logic that destroying the rainforest is bad is at best difficult to refute, Preston relies on the theme of Nature's Revenge, where Nature (or "Gaia," to use a term fashionable for many years) is possessed of human emotions and perhaps a teacher's instinct to drive a good lesson into a wayward pupil, in this case humans mucking about in places where they shouldn't go. Gaia, in effect, has got some attitude. Even more subdued tomes than Preston's uses the same intellectual framework: Laurie Garrett, in her bestseller *The Coming Plague* (1995), emphasizes a similar notion, subtitling her book "Newly Emerging Diseases in a World Out of Balance."[10] That the 2009 influenza pandemic appears to have begun on an industrialized Mexican pig farm fits in to a degree with this meme, as does the SARS outbreak, which began in rural southern China in late 2002 and was traced to exposure to the wild civet.[11] The moral here is not really that technology kills, but rather that humanity has passed some point of no return with its technology, and viruses are one of the methods by which Nature defends herself.[12]

The narrative of Nature's Revenge, however, masks a deeper, more sinister narrative: that of xenophobia. The raise-the-hair-on-the-back-of-your-neck drama of an *Outbreak* or *The Hot Zone* lies not merely in a lethal microbe "invading" the United States, but in *where* that microbe invades, to wit: a place that becomes a symbol for American purity (both in the literal sense of "uninfected," and in the metaphorical sense of morally uncorrupted), the Norman Rockwellian town where everybody knows your name. The fictional cinematic Cedar Creek appears to be a small, well-kept (read: white) northern California town. Reston, Virginia, however is a very real city where a very real Ebola outbreak occurred. Yet Preston's Reston is described with almost the same wistfulness that Cedar Creek is displayed:

> Reston was surrounded by farmland, and it still contains meadows. In spring the meadows burst into galaxies of yellow mustard flowers, and robins and thrashers sing in stands of tulip trees and white ash. The town offers handsome residential neighborhoods, good schools, parks, golf courses, excellent day care for children. There are lakes in Reston named for American naturalists (Lake Thoreau, Lake Audubon) . . .Reston is situated within easy commuting distance of downtown Washington . . .

While Reston does exist and was not simply chosen by someone as the site of a drama, it does not mean that Reston is not also an icon for a clean, decent America about to experience a siege of pure evil. The "fact" of Reston's general suburban, purified homey-ness reinforces the metaphor which drives the tension in a fictional story like *Outbreak*, which in turn fuels the American anxieties of contamination and were plainly visible in the breathless media coverage of, among other episodes, the Kikwit outbreak (Ebola, 1995),[13,14] the Uige outbreak (Marburg, 2004),[15,16] the Bundibugyo outbreak (Ebola, 2007-2008),[17] and SARS. HIV, of course, deserves a special mention, managing to provide metaphors of contamination by being both African in origin and originally transmitted by homosexual men in the late 1970s.

The violation of American purity through faceless horrors from the jungle was also evident in Steven Spielberg's *Arachnophobia* (1990). Although this half-horror, half-comedy movie featured spiders rather than viruses, its plot is almost conspiratorially similar to *Outbreak*. In this case, scientists manage to accidentally bring a killer, man-eating spider from the tropics of South America to, interestingly, a sleepy rural California town. The spider manages to find a genetic cousin with which to reproduce, resulting in hundreds or thousands of poisonous and hungry offspring, who manage to eat a few people and make a general nuisance of themselves until eliminated by the movie's protagonists, among them a white, middle-class doctor who had recently moved to the small town to avoid the evils and hassles of The Big City.

# Metaphor, the Media, and Politics

These anxieties and preoccupations had real-world consequences not only in the media coverage of the outbreaks noted above, but in the political dialogue they produced, where again the metaphor of virus-as-violator of American sanctity and purity came to the fore. In the United States, much of the early discussion of the 2009 influenza pandemic focused on the political dimensions of the U.S.-Mexican relationship, where the virus was used as a convenient prop to underline anger about immigration, legal or otherwise. For instance, Jay Severin of WTKK-FM in Boston—a typical example of the class of overheated right-wing nativists that populate talk radio—used his radio platform to describe Mexicans as, among other things, "primitives," "leeches," and "exporters of women with mustaches and VD . . . [and] swine flu."[18] Severin was suspended for his remarks, which as the saying goes, is saying something, yet he was hardly the only radio show host that waxed disgustic on the nativist theme; his remarks were merely the most outrageous, which again is saying something.

Other lethal microbial outbreaks have occurred where it is impossible, or at least difficult, to read a narrative of invasion. Indeed, while the Four Corners hantavirus outbreak (Sin Nombre, 1993)[19] did originate near the "exotic" locale of a Navajo reservation—and thus can be regarded as partially "other"—multiple strains of hantavirus have since been discovered in nearly all parts of the United States (e.g. Black Creek Canal virus, FL; Prospect Hill virus, MD; Bayou virus, LA, TX; Monongahela virus, PA; New York virus, NY), the hosts being nothing more than common field mice or meadow voles.[20] Other home-grown biological threats include plague and anthrax, with roughly ten to fifteen cases of the former and three-four cases of the latter per year.[21] Such events cannot lend themselves to the narrative of Nature's Revenge or of American invasion, and do not seem to have as tight a hold on American popular anxieties as microbes from abroad.

These narratives, of course, do affect real political relations between, say, the United States and Africa (or more broadly between the so-called First World and Third World), not to mention the manner in which international scientific and medical studies are conducted and understood. A common African "reading" of AIDS, for instance, was quite distinct from the American reading. Far from seeing themselves as the metaphorical aggressor, they regarded themselves, not without justice, as *the colonized*. Indeed, many African doctors had a reverse scenario in mind regarding the origin of AIDS, subscribing to a theory that a contaminated supply of polio vaccine, *manufactured in the United States and imported to Africa*, was the source of the virus.[22] It would turn out that this hypothesis would be wrong, but initially it was no more or less credible than the American-favored "Out Of Africa" hypothesis (which was subsequently proven, but not definitively so until 1999).[23] The various hypotheses put forth during the 1980's about the origin of AIDS tell as much about how their authors came to favor them than it does about "pure" science, and the influence of the virus-as-metaphor on scientific thinking.

Similarly, the fact that the 2009 H1N1 pandemic appears to have originated from pig farms in Mexico has not been missed in the ensuing flap about transnational agribusiness, large-scale industrialized food production, and its safety or lack thereof. Detractors of these practices were quick to interpret the early events in Mexico as Nature's Revenge—that is, that the cause of this pandemic wasn't part of a natural cycle and constant interplay between virus and mammal or fowl, but rather was due to the fact that these farms housed tens of thousands of pigs herded together in huge pens (increasing the opportunity for viral particles to genetically rearrange themselves into new forms) and required large exhaust systems to ventilate waste (increasing the opportunity to spread rearranged viral particles to humans). Yet such observations conveniently ignore the fact that the 1918 pandemic did not require large-scale industrialized agriculture, and although the origins of The Great Influenza remain unsettled, it is certainly possible that it arose from one farm in rural China where one family lived in proximity to its pigs and ducks.[24]

To think properly about the dangers of our current industrial behavior and whether that behavior really does put us at risk of lethal microbial outbreaks requires us to strip away metaphor at every turn. Whether the metaphor relates to the dangers of modernization or of fear of contagion from "the other" is irrelevant, even if each of those metaphors lead to different real-world effects. The only way by which appropriate scientific and political dialogue about lethal viruses and their impact on humans can take place is to scrutinize, in each instant, whether metaphor has contaminated the thought process, and whether different language might be better suited to describe what is taking place.

# From Imagery to Images

Of course, a central problem in the virus-as-metaphor phenomenon is that *language* is of little structural importance to the metaphor itself in the age of so-called "mass media." The lay public owes much of its understanding of "virus" to television, and television's near-total reliance on images over language to make its points can't be underestimated in its ability to mislead, intentionally or not. During the Ebola outbreaks, the TV news version of "virus" was less a thing to be discussed, much less learned about, as it was to be *seen*.

One might argue that the artistic movement of impressionism was one century ahead of its time, as the television news organizations really only succeeded in leaving the visual impression of what lethal viral outbreaks meant during the various African filoviral outbreaks in the 1990s: a compilation of B-footage showing devastation (cut to image of mass gravesites), destitution (cut to image of crowds of impoverished Africans), the Heroic Measures being taken to contain it (cut to image of health workers in "spacesuits"), and maybe a map showing the location of the terror. The message of the medium, if it can be said to have any message at all, is *feel fear*. This is a central feature of what passes for news in the mass media today, not just in relation to infectious diseases but in the specter of terrorism, the threat of climate change, rogue nukes, and all the rest.[25]

The 2009 H1N1 story, as told by the television media, started out similarly, with Mexico as a reasonable stand-in for Africa (read: not quite as exotic, nor as impoverished, nor as black, but such subtleties will hopefully be lost on the American audience). However, as the pandemic shifted out of Mexico and into the States, the preoccupations turned to domestic "preparedness" and the boob-tube images turned to B-footage of labs preparing a vaccine, long lines outside doctor's offices, and the perfunctory image of the Centers for Disease Control headquarters in Atlanta. The change in the visual texture might be regarded as moving from the message of *feel fear* to *the fear is here!*

# Conclusion

Particularly in the age of television, the ability of reporters to provide the general public with stories about viruses devoid of metaphor remains very much an open question. As the internet and the nationwide availability of high-quality newspapers has developed, it is possible that a two-tiered system of news may develop, with television news providing the superficial analysis and Internet sites supplying substance, although the fluid nature of the Internet makes such predictions difficult. (Obviously, a "two-tier system" has long been in place and television has always on average been less substantive than quality print media. The difference over the past ten to fifteen years is that, if one lived in a small town in, say, central Ohio in the mid-1980s, the best print daily they could read would have been Cleveland's *The Plain Dealer*, which though a fine paper, did not have the breadth or depth of *The New York Times*, which can now be easily found in such remote places.)

Whether through television, books, movies, or Internet blog entries, the various accounts of lethal viral outbreaks, whether fictional or not, influence other stories, less visible stories, which have real consequences in a political arena: a "hot zone" including multinational corporations reaping profits from exploiting the resources of the third world in general (and Africa in particular); the governments of the former Soviet Union and the United States fighting proxy wars in such countries; women in these places who are forced into prostitution to survive, becoming unwitting amplifiers of disease; and citizens of all republics who are affected by the spread of whatever disease happens to have capitalized on these and other factors. These narratives are constantly intermingling—contaminating—one another, and perhaps like Nature's warnings, we ignore that dialogue at our peril.

# Notes

1. Final estimate of the Aggregate Hospitalization and Death Reporting Activity (AHDRA) of the Centers for Disease Control, August 30, 2009-April 3, 2010. The total number of H1N1 laboratory–confirmed hospitalizations during this period was 41,914. While these numbers likely represent an underestimate, in terms of mortality, the novel H1N1 pandemic was remarkably mild regardless. Statistics available at www.cdc.gov/flu/weekly (accessed July 10, 2010).

2. No discussion about virus-as-metaphor would be complete without a nod to Susan Sontag's seminal works *Illness as Metaphor* and, even more importantly in this discussion, *AIDS and Its Metaphors*. I am in particular indebted to Sontag's "AIDS" for laying out the framework for thinking about the motif of "other" in relation to disease, especially disease transmitted by viruses. Sontag S, *Illness as Metaphor and AIDS and Its Metaphors*. (New York: Picador (Macmillan), 2001).

3. "No Yellow Fever in Frankfurt," *New York Times,* August 26, 1967, 4 (63 words). "2 More Germans Die From Monkey Virus," *New York Times*, September 5, 1967 (173 words).

4. The convention of naming viruses after the communities in which the initial outbreak occurs dates to the first half of the twentieth century, when the virus causing Rift Valley Fever (clinically described in 1915) was isolated in 1931. Other early outbreaks with identified viral agents included Crimean Hemorrhagic Fever Virus (1944; now called Crimean–Congo Hemorrhagic Fever Virus) and Omsk Hemorrhagic Fever Virus (1945). The virus causing "Korean Hemorrhagic Fever," identified by researchers during the Korean War, became renamed "Hantaan virus" after a South Korean river near the site of outbreaks. In general, locals aren't precisely thrilled when their village or region is used to name a deadly microbe, and that holds particularly true when those places are peopled by groups traditionally marginalized by past European or American governments. The practice came to a head in the United States just about two decades ago, when a very deadly virus in the Four Corners area of the United States infected a small number of people. A new type of bunyavirus, the researchers studying it were prepared to give it the place name as per protocol, until a group of local protesters, including many members of the Navajo nation, made it clear that they did not want their community disrespected this way. As a consequence—and a compromise—this virus today literally has "no name": it is known as the Sin Nombre Virus.

5. Murphy, Kiley, Fisher-Hoch, "Filoviridae: Marburg and Ebola viruses." In *Virology*, ed. Fields BN, Knipe DM (New York: Raven Press, Ltd., 1990), 933–42.

6. The original Marburg outbreak was mild even by subsequent Marburg outbreak standards: the Uige outbreak in 2004 in Angola, at 252 documented infections the largest Marburg outbreak to date, had a 90 percent mortality rate. For these data, see: http://www.cdc.gov/ncidod/dvrd/spb/mnpages/dispages/marburg/marburgtable.htm (accessed July 10, 2010).

7. U.S. Gross Domestic Takings of ~$67 million, and an additional International Takings of ~$120 million. http://www.imdb.com/title/tt0114069/business (accessed July 10, 2010).

8. Preston, *The Hot Zone: A Terrifying True Story* (New York; Anchor, 1995).

9. Jahrling, Geisbert, Dalgard, et al., "Isolation of Ebola virus from imported monkeys in the United States." *Lancet* 335 (1990): 502-5.

10. Garrett, *The Coming Plague: Newly Emerging Diseases in a World Out of Balance* (New York: Farrar, Straus and Giroux, 1994).

11. Zhao, "SARS molecular epidemiology: a Chinese fairy tale of controlling an emerging zoonotic disease in the genomics era." *Philos Trans R Soc Lond B Biol Sci.*; 362, no.1482 (June 29, 2007): 1063–1081.

12. We could, for instance, include the Toxic Shock Syndrome outbreak (1978-81) and its relation to the poorly designed *Rely* tampons by Procter & Gamble. Though the outbreak did not occur from making incursion into the rainforest, "the story" of TSS can also be read as Nature's Revenge. However, TSS is caused by a bacterial rather than viral infection, so that would be cheating.

13. Centers for Disease Control and Prevention, "Outbreak of Ebola viral hemorrhagic fever—Zaire, 1995." *Morbidity and Mortality Weekly Report* 44, no. 19 (May 19, 1995): 381-2.

14. Contreras, "On Scene in the Hot Zone." *Newsweek*. May 29, 1995.

15. Centers for Disease Control and Prevention. "Outbreak of Marburg virus hemorrhagic fever—Angola, October 1, 2004-March 29, 2005," *Morbidity and Mortality Weekly Report*, 54 no. 12 (Apr 1, 2005): 308-9.

16. British Broadcasting Corporation (BBC) News. "Angola's town of fear and lonely death." May 3, 2005.

17. Towner, Sealy, and Khristova et al., "Newly discovered ebola virus." *PLoS Pathog* 11 no. e1000212 (Nov 4, 2008). Epub Nov 21, 2008.

18. Abel, "Severin suspended for comments about Mexican immigrants." *The Boston Globe*, May 1, 2009.

19. Centers for Disease Control and Prevention. "Outbreak of Acute Illness—Southwestern United States, 1993." *Morbidity and Mortality Weekly Report* 42, no. 22 (June 11, 1993): 421-4.

20. Khaiboullina, Morzunov, St Jeor, "Hantaviruses: molecular biology, evolution, and pathogenesis." *Curr Mol Med* 5, no. 8 (Dec. 2005): 773-90.

21. Again, these are bacteria rather than viruses, but why quibble? The anthrax I am referring to does not include the bioterrorist anthrax episode of 2001; anthrax is naturally acquired usually through working with the hides of animals, often when making drums or furniture bearing animal hides, http://emergency.cdc.gov/agent/anthrax/faq/pelt.asp (accessed July 10, 2010). Data on anthrax and plague from information pages at http://www.cdc.gov (accessed July 10, 2010)

22. Discussed in Worobey, Santiago, and Keele et al., "Origin of AIDS: contaminated polio vaccine theory refuted," *Nature* 428, no. 6985 (Apr 22, 2004): 820.

23. Gao, Bailes, and Robertson, "Origin of HIV-1 in the chimpanzee Pan troglodytes troglodytes," *Nature* 397, no. 6718 (Feb 4, 1999): 436-41.

24. A fine discussion of the origins of the 1918 pandemic, including the theory that the great plains of the United States was the site of The Great Influenza's beginnings, can be found in Barry J, *The Great Influenza: The Epic Story of the Deadliest Plague in Histor* (New York: Viking, 2004).

25. For a complete discussion, see Siegel, *False Alarm: The Truth About the Epidemic of Fear* (Hoboken: Wiley, 2005).

# Chapter 7

# Gender, Germs, and Dirt

## Sharyn Clough

Within philosophy of science in the analytic tradition, and in the public understanding of science more generally, there has been a tendency to view any political interest as a source of bias and error in scientific research.[1] As feminist studies of science throughout the 1980s and 1990s have documented, there have indeed been many cases where political interests, such as sexism and racism, have introduced bias and errors, negatively affected the empirical adequacy of scientific studies.[2]

However, as feminist and other progressive scientists and philosophers of science have worked to address sexism and racism in science, they have found themselves in the awkward epistemic position of criticizing these political interests, while simultaneously offering prescriptions for better scientific research—prescriptions that are themselves explicitly aligned with yet another (e.g., feminist) set of political interests. What has become clear, though, is that in a number of well-documented cases, feminist-informed prescriptions for science have made empirical improvements over sexist research.[3] Can it be that while some political interests simply bias scientific research, some other political interests (such as feminist interests) can be used as effective resources for increasing the empirical adequacy of research? And if so, how? Much work in feminist science studies has aimed at grappling with these questions, beginning, of course, with Harding.[4] In this chapter I discuss a compelling set of epidemiological and immunological studies that inform the "hygiene hypothesis" and argue that this hypothesis is made more empirically robust by augmenting it with a particular set of feminist political interests.

As will be discussed in more depth, below, the hygiene hypothesis was developed to explain the correlation between increased sanitation, and increased incidence of allergies, asthma, and other auto-immune diseases. Especially in the industrialized nations of the North and West, it is hypothesized, our lower rates of exposure to certain kinds of bacteria and other micro-organisms, especially in childhood, have had unintended negative consequences for our immune health as adults.

The one common denominator that has received no critical attention by hygiene hypothesis researchers however, is that, in the industrialized North and West, women are over-represented in all the relevant clinical populations.

Women have higher rates than men of asthma, allergies, IBD, Crohn's, rheuma-toid arthritis, multiple-sclerosis, Grave's disease, and Lupus, to name a few of the relevant auto-immune disorders.[5] The "feminization" of these diagnostic cat-egories has been well-acknowledged by clinicians who treat and study these ill-nesses, but not by the immunologists and epidemiologists involved in studying the hygiene hypothesis.

Despite the fact that the clinicians who treat and study these illnesses have noticed the sex differences in the populations they treat, their own explanatory lapse has come in the form of a trend towards reductionistic biomedical explana-tions for these differences that put the focus on physiological, hormonal, and ge-netic accounts, with little to no attention paid to the ways these same processes are affected by complex environmental factors, such as patterns of hygiene and sanitation, or social factors, such as the interweaving of the effects of gender, race, and economic hierarchies.[6] It is not surprising then that reductionistic bio-medical accounts of the differences in the relevant morbidity rates between women and men continue to leave a significant amount of variation unexplained.

This is precisely the explanatory gap that the hygiene hypothesis is suited to address, especially when a feminist understanding of the gendered socialization of children is added to the epidemiological and immunological picture. That standards of cleanliness are generally higher for girls than boys, especially under the age of five when children are more likely to be under close adult supervi-sion, is a robust phenomenon in industrialized nations, and some studies point to a cross-cultural pattern.

My use of cautionary phrases such as "generally higher for girls than boys," is meant to indicate a respect for the complexity of gender categories as they are interwoven in the hygiene story with categories of ethnicity and socio-economic status. While I attend to these latter categories, at least in terms of nationality and urbanization, more analysis on these fronts is needed to fill in the explanato-ry picture.

In what follows, I review the case for the hygiene hypothesis.[7] I then dis-cuss the feminist political commitments that make the link between gender and the hygiene hypothesis salient, and present feminist-informed sociological and anthropological research to argue that the feminine gender-role socialization of girls includes higher standards of cleanliness than does the masculine gender-role socialization of boys. I argue that, insofar as the hygiene hypothesis suc-cessfully identifies standards of hygiene and sanitation as mediators of immune health, then, properly augmented by feminist analyses of the gendered standards of cleanliness in children, the hypothesis can account for the unexplained varia-tion in the relevant morbidity rates between men and women.

Finally, I argue that by making the link between gender and hygiene visible, these feminist political commitments do not bias the immunological and epide-miological research. Instead, insofar as research supports the explanatory role of the broadly environmental and social factors underlying the hygiene hypothesis, the political commitments I prescribe have the effect of *increasing* the empirical adequacy of that research, specifically by reconceiving of relevant sources of ev-idence, and opening up further avenues for study.

# The Hygiene Hypothesis

## Germs, Allergies, and Asthma

Industrialized nations of the North and West have experienced increasing rates of asthma and allergies.[8] This increase has been explained by appealing to increased standards and practices of hygiene and sanitation. Experimental support with mice has shown that a variety of allergic responses can be decreased via exposure to the bacteria *M. Vaccae*.[9] Support for the hypothesis also comes from studies of human populations that document the protective effects of farm environments for children.[10] Compared to children raised in urban settings, children raised on farms have lower rates of allergic rhinitis and/or conjunctivitis. Kilpeläinen, *et. al.*, conclude that "environmental exposure to immune modulating agents, such as environmental mycobacteria could explain the finding."[11] Exposure to two or more domestic pets has been shown to have a similar protective effect: "exposure to two or more dogs or cats in the first year of life was associated with a lower prevalence of allergic sensitization at age six-seven years."[12] A more recent study of children in the Philippines found that increased exposure to microbes in childhood predicted increased immune health in adulthood.[13]

## Parasites and Inflammatory Diseases

Elliott, Summers, and Weinstock have theorized that the increase in other immune-system malfunctions such as inflammatory-bowel disease (IBD) and Crohn's disease may also be related to the hygiene hypothesis.[14] They observed that in contemporary urban environments where humans are largely free of contamination by parasitic worms, rates of these sorts of diseases have increased dramatically. In Israel, for example, Zvidi, *et. al.* report that "the prevalence rate rose from 25.53/100,000 in 1987 to 65.11/100,000 in 1997, and then to 112.99 in 2007."[15] Since some parasitic worms seem to have a "calming" effect on the immune system, it seems likely that the trends are related. Indeed, clinical trials have shown that exposure to the eggs of the *Trichuris suis* whip worm can reduce the severity of symptoms in patients with Crohn's.[16]

# Feminist Political Interests and the Hygiene Hypothesis

The over-representation of women in the clinical populations discussed above is largely ignored in the current literature, though the pattern can be accounted for by my augmented hygiene hypothesis regarding the differential gender socialization of children. However, the strength of the augmented hypothesis is only obvious against a backdrop of feminist political commitment, and it is to the details of that commitment that I now turn.

Very generally, the feminist politics of relevance concern the project of documenting, deconstructing, and ameliorating the varying social pressures that inform what it means to be a gendered body. Within this feminist political project is embedded a cluster of views that are at once both descriptive and prescriptive, e.g., the view that the content of the social roles assigned to boys and girls, men and women, is significantly driven by deeply held cultural commitments that are in some important sense: a) arbitrarily assigned relative to features such as secondary sex characteristics; and b) vigorously, though often unconsciously, enforced and rewarded from a very young age.

While this latter feature means that our current assignment of social roles is not easily modified, most feminists believe that the historical evidence regarding human flourishing shows us that there are more relevant criteria for assigning social roles, such as individual interest and/or skill. Indeed, when girls are given the same chances, training, and encouragement as boys, they seem to be just as capable as boys in a variety of tasks not often associated with femininity; the pattern holds similarly for boys and tasks not often associated with masculinity.

Social scientists informed by feminism have amassed a great deal of data supporting this cluster of political views, but there are of course a number of people who do not share these political interests, believing instead that gender role assignments are less than arbitrary, perhaps "natural" or mandated by theological design, and, that, hence, individual interest and skill are less relevant. When one holds these latter views, or is unaware that there is a position to be taken on these points, or has heard of the feminist positions outlined, and thinks them reasonable, but does not identify them as key organizing principles, then one is less likely to notice, or think necessary or even relevant, an analysis of the way in which gender roles can have an effect on any particular scientific hypothesis. Gender, in these instances, is easily "disappeared," to use Helen Longino's phrase.

Of particular relevance to my analysis of the hygiene hypothesis is the feminist claim that masculine gender-role assignment, broadly construed, involves a social acceptance of playing in dirt and mud for those (typically boys) so assigned, an acceptance that does not extend to the feminine gender role, broadly construed (and typically assigned to girls). These differential social expectations regarding cleanliness are reflected in and reinforced by gender differences in children's clothing, participation in sports, and adult supervision of children's play.

## Gendered Norms of Cleanliness

Feminist-informed sociologists in the industrialized North and West have documented that girls are dressed more often than boys in clothing that is not supposed to get dirty, and that restricts the sorts of movements that would get one dirty in the first place.[17] Girls do not participate in sports with the same frequency as boys, and girls more often than boys play indoors.[18] Insofar as many sports, and outdoor play generally, the chances for exposure to the microorganisms found in dirt (and there is abundant evidence to support this), then boys will have greater rates of exposure to these micro-organisms than will girls.[19] Finally, parents structure and supervise the play of girls more than that of boys, which is likely to result in girls being kept cleaner than boys.[20]

Many parents continue to reinforce traditional gendered norms of hygiene in their preschool children, as expressed, for example, in how children are dressed. In a study of American children in a preschool setting, one third of the five-year-old girls came to school in dresses *each day*.[21] Of relevance to the question of cleanliness, Martin noted that being in a dress limited the girls' "physicality." However, she added, "it is not only the dress itself, but knowledge about how to behave in a dress that is restrictive. Many girls already knew that some behaviors were not allowed in a dress. This knowledge probably comes from the families who dress their girls in dresses."[22] Not surprising, given that girls receive more parental supervision and direction regarding cleanliness than do boys, girls more than boys are taught to police themselves—to be vigilant about their appearance and cleanliness. One particular observation of five-year-old girls in Martin's study is worth quoting at length:

> Vicki, wearing leggings and a dress-like shirt, is leaning over the desk to look into a "tunnel" that some other kids have built. As she leans, her dress/shirt rides up exposing her back. Jennifer (another child) walks by Vicki and as she does she pulls Vicki's shirt back over her bare skin and gives it a pat to keep it in place. It looks very much like something one's mother might do.[23]

These young children have already internalized the rule that when wearing a dress, even a dress-like tunic with leggings, they must constantly monitor their decorum—who knows what immodesty might otherwise result, what dirt (metaphorically and literally) might cling.

Sociologist Thorne documents gender norms in children's play at the elementary school level in her book, *Gender Play*.[24] Of particular relevance is her discussion of "cooties" and other "pollution rituals"—concerns that are especially prevalent, she notes, in children ages six to nine. Her observations, taken from field work in the late 1970s and early 1980s at schools in Michigan and California, show that girls are far more likely than boys to be associated with cooties and to be ostracized as carriers of cooties.[25] The clear message is that, unlike boys, girls need to guard against these and other forms of pollution. "Girls as a group are treated as an ultimate source of contamination.[26] While individual boys are sometimes also so marked, in these cases, she notes, it is the boy's ethnicity or physical ability that is used to set him up as a source of pollution, rather than his gender, *per se*. She also notes a common pattern in this research, namely that boys, more often than girls, played outdoors.[27]

# Increasing the Empirical Adequacy of the Hygiene Hypothesis

Recall that I began with the general question regarding whether and how it is that while some political interests can bias scientific research, other political interests (such as some feminist interests) can empirically strengthen research. Particular interests and particular case studies are needed to defend this claim, and as promised, I began by identifying a particular case study, the hygiene hypothesis, and the particular feminist political interests that strengthen that hypothesis. The next feature of my project is to address in more detail the improvements to the hygiene hypothesis that attention to feminist interests allow.

Beginning with the case of asthma, age interacts with sex in a way that is consistent with my augmented hygiene hypothesis regarding the gender socialization of children. Before puberty, boys have higher rates of asthma than girls.[28] After puberty, the sex difference reverses, with women having higher rates than men. Osman reports that the reasons for the age-link remain unclear. As mentioned above, there are certainly a number of competing biomedical explanations for the "over-active" immune systems of women relative to men,[29] however, there are no accepted biomedical explanations available for the over-active immune systems of boys relative to girls. If my augmented hypothesis is right, it might be that there is a critical period involved, a developmental period during which the immune system, properly exposed to potential allergens, responds with asthmatic symptoms, and after which shows a "settling effect." Those children, typically boys, properly exposed during the critical period, respond with asthmatic symptoms early on, but then their symptoms abate. Those children, typically girls, not exposed during the critical period, respond with asthmatic symptoms later, and for the rest of their lives.

In fact, a critical period of just this sort was found in Ownby, et. al., the study referred to above, that showed that having two or more pets in the home at infancy protected children against allergies at six years of age.[30] The positive effect was not found if the pets were introduced later than infancy. And, in what the authors note as a "puzzling" aside, the protective effect of pets in the home was significantly more marked for boys than girls.[31]

Thinking in terms of gender differences in hygiene also helps identify new relevant sources of evidence for the hygiene hypothesis and opens up further avenues for study. While epidemiologists and immunologists have not yet linked gendered norms of cleanliness to morbidity rates for auto-immune disorders, there are some epidemiological studies of children that mention gender differences in exposure to dust, dirt, and germs and these studies, not previously believed to be relevant, could be used to provide support for and point towards the further development of the hygiene hypothesis.

Eating dirt, or geophagia, is a very reliable way to ingest micro-organisms, and in a study from rural Guinea that examined the ingestion of parasites via geophagia, boys under the age of ten are significantly more likely than girls to be infected by these parasites.[32] Consider also studies of the transmission of *Ascaris lumbricoides* (a harmful intestinal roundworm) among rural populations in

Southern Ethiopia.[33] The transmission route typically involves "ingestion of infective eggs from soil contaminated with human feces or uncooked vegetables contaminated with soil containing infective eggs."[34] In Southern Ethiopia, homes typically have dirt floors, infants are often accompanied by domestic dogs throughout the day, and livestock are brought into homes at night. The dogs and livestock as well as the dirt floors are all sources of the eggs. Children are the primary victims of roundworm infection.[35] At one clinic, "70 percent of all outpatients treated for helminthiases [intestinal parasites] were children under fourteen years of age,"[36] with infection rates highest among one to four-year-olds. Of note here is the sex difference: 20.5 percent of males in this age group had *Ascariasis* infection, as opposed to only 13.5 percent of females. The researchers remarked that this gap closes in later ages, e.g., five to fourteen-year-olds, where rates lower—in males to 8.9 percent and females to 11.3 percent—"making it difficult to establish statistically significant sex-differences in worm infection."[37]

However, the change across age groups might be explained by differential hygiene expectations for boys and girls (more on the cross-cultural strength of these expectations below). Infants, both male and female, are likely to be under greater parental supervision than are older children, and for infants gendered feminine rather than masculine, greater supervision is likely to come with increased restrictions on how and where they play. Perhaps these gendered facets of parental supervision explain the fact that in the one to four age group, significantly fewer girls than boys were found to have parasites. In addition, the reported decrease in parasitic load with age might be similar to the decrease in asthma rates with age that is reported in boys and girls in countries in the industrialized North and West. Again, a critical period for exposure to the relevant parasites might be at work, disadvantaging those children, typically girls, who have fewer opportunities for exposure.

If the modalities of exposure to harmful micro-organisms such as the *Ascaris lumbricoides* parasite are similar to those for the more helpful micro-organisms that calm over-active immune systems (such as *M. Vaccae*, and the *Trichuris suis* whip worm) then epidemiological studies like these, attending to gendered norms of cleanliness, could serve as further sources of evidence in support of the hygiene hypothesis. If the augmented hygiene hypothesis is true, that is, if increased hygiene negatively affects immune health, and immune health differs by sex (as research shows it does), then we should see sex differences in the morbidity rates for these illnesses. And here we do.

The epidemiological research in rural Guinea, and Southern Ethiopia, cited above, introduces another avenue of study that could provide evidence for the hygiene hypothesis. Such research undertaken outside the industrialized North and West provides the opportunity for a number of cross-cultural, natural experiments.

The first sort of experiment would evaluate whether the gender norms that place higher standards of cleanliness on girls than boys hold across different cultures. Some of the sociological and anthropological research on gendered norms of cleanliness in nonindustrialized nations of the global South and East suggests that this is the case.

In her field work in Bengali, India, anthropologist Lamb found that women and girls are expected to bathe more often than men and boys—expectations that are related to views of women and girls as naturally dirtier than men and boys.[38] For these women and girls, the practice of bathing, often twice or more times daily, consists mostly in a ritual rinsing with water, rather than wringing one's hands with anti-bacterial soap, or guarding against cooties, but the gender differences in cleanliness here clearly run parallel to the purity notions associated with femininity in the North and West.

In the Caribbean country of Guyana, sociological research on parental socialization preferences showed that parents rated neatness and cleanliness as "more desirable for girls [than boys] in all age groups."[39]

A second kind of natural experiment involves examining whether those epidemiological studies of the rural South and East that reported higher levels of ingestion of micro-organisms in boys, also found this to be correlated with those boys having a lower incidence of allergies, asthma, and IBD. We already know that the incidence and prevalence of these diseases is lower in the more rural nations of the South and East, relative to the industrialized North and West, but we don't know whether morbidity patterns in the South and East feature the same sex differences as are found in the North and West. Insofar as the gendered socialization patterns that have been well-identified in the industrialized North and West continue in the nonindustrialized, primarily rural settings of the South and East, we can expect that there will be sex differences in the relevant morbidity rates.

One study comes close to confirming this expectation. Researching allergies among the Hiwi settlements of Venezuela, anthropologists Hurtado et. al. noted that, consistent with the hygiene hypothesis, these populations had lower rates of allergies than are typically found in populations of the industrialized North and West.[40] They also noted that Hiwi girls spend significantly more time than do boys engaging in "grooming behaviors," and that these behaviors "serve to eliminate ectoparasites."[41] What they did not note in their study was whether there were any sex differences in parasite exposure between Hiwi boys and girls, though the grooming behavior suggests there is. They also did not note whether there were any sex differences in incidence and prevalence of allergies in either children or adults. However the presentation of their data suggests that they have this information available (they present data comparing Hiwi girls and women to girls and women from Western populations, for example). In the absence of attention to the socialization processes that differentially affect hygiene expectations for boys and girls, it is likely that the researchers did not think that sex differences within the Hiwi populations were relevant for analysis and presentation. Paying attention to gender differences in hygiene provides a means of recognizing potential evidence for the hygiene hypothesis that might be otherwise ignored.

# Conclusion

One of the ways that feminist and other progressive philosophers of science have tackled the problem of bias in science has been to document the frequency with which a variety of political interests present in early stages of the discovery of a scientific hypothesis can, and do, go on to bias hypothesis testing and results.[42] It seems clear from their careful documentation that political interests are ubiquitous in all aspects of scientific activity. But it should also be clear by now that not all political interests are of a piece. As I have argued, some political interests can actually increase the empirical adequacy of scientific research in a variety of ways.

I have argued that hygiene hypothesis researchers have not sufficiently attended to, or accounted for, the "feminization" of the morbidity rates they seek to explain, a lapse that is especially problematic as these sex differences are very well-documented in nations of the industrialized North and West. While clinicians who treat and study these illnesses *have* noticed sex differences in the populations they treat, they have tended towards reductionistic biomedical explanations for these differences that put the focus on physiological, hormonal, and genetic accounts. This is a second kind of lapse that has resulted in little to no attention being paid to the ways these same physiological, hormonal, and genetic phenomena are affected by complex environmental factors, such as patterns of hygiene and sanitation, or social factors, such as the interweaving of the effects of gender, race, and economic hierarchies.

As it stands, the hygiene hypothesis already enriches and indeed moves beyond reductionistic biomedical approaches to immunological illnesses, by attending to complex environmental and social factors such as sanitation and hygiene standards. I have argued that, insofar as the hygiene hypothesis successfully identifies standards of hygiene and sanitation as mediators of immune health, then, introducing a feminist analyses of the gendered standards of cleanliness in children adds important empirical resources to the explanatory picture.

Having conducted an interdisciplinary review of feminist research on the gender socialization of children, I have shown that, insofar as social preferences for cleanliness are enforced more aggressively for girls than boys, this gender difference leaves girls with lower rates of exposure than boys, to an array of micro-organisms. Attention to differences within gender categories, such as nationality, and urbanization, make clear that the gendered pattern is robust.

The feminist political interests that inform the augmented hypothesis help fill in some of the explanatory gaps in our current understanding of why it is that, in industrialized nations of the North and West, at least, women are more likely than men to suffer from auto-immune diseases. These political interests also respond to a number of outstanding puzzles in the hygiene hypothesis research, make available new sources of evidence, and suggest designs for a number of cross-cultural and other natural experiments.

# Notes

1. See, for example, the public views surveyed in Steel, List, Lach, and Shindler, "The role of scientists in the environmental policy process: A case study from the American west." *Environmental Science & Policy* 7 (2004): 1–13.

2. Fausto-Sterling, *Myths of Gender: Biological Theories About Women and Men*, 2nd ed. (New York: Basic Books, 1992); Schiebinger, *The Mind Has No Sex? Women and the Origin of Modern Science* (Cambridge: Harvard University Press, 1989); Tavris, *The Mismeasure of Woman*. (New York: Simon and Shuster, 1992); Harding, "Rethinking Standpoint Epistemology: What is 'Strong Objectivity'?" Pp. 49-82 in *Feminist Epistemologies*, edited by Alcoff and Potter. (NY and London: Routledge, 1993); and Spanier, *Im/partial Science: Gender Ideology in Molecular Biology*. (Bloomington: Indiana University Press, 1995).

3. Fausto-Sterling, *Myths of Gender: Biological Theories About Women and Men*, 2nd ed. (New York: Basic Books, 1992); Anderson, "Uses of value judgments in science." *Hypatia* 19, no. 1 (2004): 1-24.

4. The original work on this topic is Harding, *The Science Question in Feminism*. (Ithaca: Cornell University Press, 1986). For more recent reviews of this work, see also Intemann, "Science and Values; Are Moral Judgments Always Irrelevant to the Justification of Scientific Claims?" *Philosophy of Science* 68, no. 3 (2001): S506-18; Clough, *Beyond Epistemology: A Pragmatist Approach to Feminist Science Studies*. (Lanham: Rowman and Littlefield, 2003); and Wylie and Hankinson-Nelson, "Coming to Terms with the Values of Science: Insights from Feminist Science Scholarship." Pp. 58-86 in *Value-Free Science? Ideals and Illusions*, edited by Kincaid, Dupré, and Wylie. (New York: Oxford University Press, 2007).

5. Jacobson, Gange, Rose, and Graham, "Epidemiology and estimated population burden of selected autoimmune diseases in the United States." *Clin Immunol Immunopathol* 84, no. 3 (1997): 223-243; Walsh and Rau, "Autoimmune diseases: a leading cause of death among young and middle-aged women in the United States." *Am J Public Health* 90, no. 9 (2000): 1463-1466; Bird and Rieker, *Gender and health: The effects of constrained choices and social policies* (Cambridge: Cambridge University Press, 2008).

6. For a critical discussion of the trend towards the reductionistic biomedicalization of these diagnoses, see Bird and Rieker, *Gender and health: The effects of constrained choices and social policies*. (Cambridge: Cambridge University Press, 2008); Epstein, *Inclusion: The politics of difference in medical research* (Chicago: University of Chicago Press, 2007); and Barker, *The Fibromyalgia Story: Medical Authority and Women's Worlds of Pain*. (Philadelphia: Temple University Press, 2005), (esp. ch. 2).

7. For a full account see Clough, in press.

8. E.g., Maziak Behrens, Brasky, Duhme, Rzehak, Weiland, and Keil, "Are asthma and allergies in children and adolescents increasing? Results from ISAAC phase I and phase III surveys in Munster, Germany." *Allergy* 58 (2003): 572–579.

9. Zuany-Amorim, et al., "Suppression of airway eosinophilia by killed Mycobacterium vaccae-induced allergen-specific regulatory T-cells." *Nat. Med.* 8 (2002): 625–629.

10. Kilpeläinen, Terho, Helenius, and Koskenvuo, "Farm environment in childhood prevents the development of allergies." *Clinical and Experimental Allergy* 30, no. 2 (2000): 201-208.

11. Kilpeläinen, et al., "Farm environment," 201.

12. Ownby, Johnson, and Peterson, "Exposure to dogs and cats in the first year of life and risk of allergic sensitization at 6 to 7 years of age." *JAMA* 288, no. 8 (2002): 969.

13. McDade, Rutherford, Adair, and Kuzawa, "Early Origins of Inflammation: Microbial Exposures in Infance Predice Lower Levels of C-Reactive Protein in Adulthood."

*Proceedings    of    The    Royal    Society    B:    Biological    Sciences.*
http://rspb.royalsocietypublishing.org/content/early/2009/12/08/rspb.2009.1795.full   (accessed July 10, 2010).

14. Summers, Elliott, Urban Jr, Thompson, and Weinstock, "Trichuris suis therapy in Crohn's disease." *Gut* 54, no. 1 (2005): 87-90.

15. Zvidi, Hazazi, Birkenfeld, and Niv, "The prevalence of Crohn's disease in Israel: A 20-year survey." *Digestive Diseases and Sciences* 54, no. 4 (2008): 848-852.

16. E.g., Summers, et al., "Trichuris suis therapy," 457-64; Erb, "Can helminths or helminth-derived products be used in humans to prevent or treat allergic diseases?" *Trends in Immunology* 30, no. 2 (2009): 75-82.

17. Martin, "Becoming a gendered body: Practices of preschools." *American Sociological Review* 63, no. 4 (1998): 494-511.

18. Pomerleau, Bolduc, Malcuit, and Cossette, "Pink or blue: Environmental gender stereotypes in the first two years of life." *Sex Roles* 22, no. 5-6 (1990): 359-367.

19. E.g. Lanphear and Roghmann, "Pathways of lead exposure in urban children." *Environmental Research* 74 (1997): 67–73.

20. Caldera, Huston, and O'Brien, "Social interactions and play patterns of parents and toddlers with feminine, masculine, and neutral toys." *Child Development* 60 (1989): 70–76.

21. Martin, "Becoming a gendered body: Practices of preschools." *American Sociological Review* 63, no. 4 (1998): 494-511.

22. Martin, "Becoming a gendered body," 498.

23. Martin, "Becoming a gendered body," 498.

24. Thorne, *Gender Play: Girls and Boys in School.* (New Brunswick, NJ: Rutgers University Press, 1993).

25. Thorne, *Gender Play*, 73-75.

26. Thorne, *Gender Play*, 74.

27. Thorne, *Gender Play* , 91.

28. Osman, "Therapeutic implications of sex differences in asthma and atopy: Community child health, public health, and epidemiology." *Archives of Disease in Childhood* 88, no. 7 (2003): 587-590; Johnson, Peterson, and Ownby, "Gender Differences in Total and Allergen-specific Immunoglobulin E (IgE) Concentrations in a Population-based Cohort from Birth to Age Four Years." *Am. J. Epidemiology* 147(1998): 1145-1152.

29. For a critical discussion of the sexism of some of these explanations see Howes, "Maternal Agency and the Immunological Paradox of Pregnancy." Pp. 179-198 in *Establishing Medical Reality: Essays in the Metaphysics and Epistemology of Biomedical Science*, edited by Kincaid and McKitrick, (Heidelberg: Springer Netherlands, 2007); Bird and Rieker, *Gender and health: The effects of constrained choices and social policies.* (Cambridge: Cambridge University Press, 2008).

30. Ownby, Johnson, and Peterson. "Exposure to dogs and cats in the first year of life and risk of allergic sensitization at 6 to 7 years of age." *JAMA* 288, no. 8 (2002): 963-972.

31. Ownby, Johnson, and Peterson, "Exposure to dogs and cats in the first year of life and risk of allergic sensitization at 6 to 7 years of age." *JAMA* 288, no. 8 (2002): 970.

32. Glickman, Camara, Glickman, and McCabe, "Nematode Intestinal Parasites of Children in Rural Guinea, Africa: Prevalence and Relationship to Geophagia." *International Journal of Epidemiology* 28 (1999): 169-174.

33. Vechiatto, "'Digestive Worms': Ethnomedical Approaches to Intestinal Parasitism in Southern Ethiopia." Pp. 241-266 in *The Anthropology of Infectious Disease: Inter-*

*national Health Perspectives*, edited by Marcia Inhorn and Peter Brown. (Amsterdam: Gordon and Breach Publishers, 1997).

34. Vechiatto, "'Digestive Worms" 241.

35. Vechiatto, "'Digestive Worms" 245.

36. Vechiatto, "'Digestive Worms" 245.

37. Vechiatto, "'Digestive Worms" 246.

38. Lamb, "The politics of dirt and gender: Body techniques in Bengali India." Pp. 213 in *Dirt, Undress and Difference. Critical Perspectives on the Body's Surface*, edited by Masquelier. (Bloomington: Indiana University Press, 2005).

39. Wilson, Wilson, and Berkeley-Caines 2003, 217.

40. Hurtado, Hill, de Hurtado, and Rodriguez, "The evolutionary context of chronic allergic conditions: The Hiwi of Venezuela." *Human Nature* 8, no.1 (1997): 51-75.

41. Hurtado, et al., "The evolutionary context of chronic allergic conditions," 63.

42. E.g., Okruhlik, "Gender and the Biological Sciences." *Biology and Society Canadian Journal of Philosophy Supplementary* 20 (1994): 21-42; Longino, "Comments on Science and Social Responsibility: A Role for Philosophy of Science?" *Philosophy of Science* 64, no. 4 (1997): 179.

# SECTION 2: THE OTHER SIDE(S) OF SCIENCE

# Chapter 8

# The Agnostic Scientist: The Supernatural and the Open-Ended Nature of Science

## Brian L. Keeley

Does the scientist *qua* scientist have a mandated position on the existence of God, gods, or the otherwise allegedly supernatural?[1] Is the appropriate attitude of the scientist *qua* scientist one of theism, atheism, or agnosticism with respect to the supernatural? It is my thesis that the appropriate "scientific attitude" is agnosticism.

While science and religious agnosticism once went hand-in-hand, in recent years, with the resurgence of arguments in support of religious atheism, the notion that the *natural* attitude of the scientist is atheistic has risen in plausibility. Eminent evolutionary biologist, Richard Dawkins, has argued forcefully that what he calls the "God hypothesis" is testable and what's more, it has been falsified.[2] A philosopher of a strong scientific bent, Dan Dennett has argued that religion and religious belief should be the subject of scientific study. Further, like Dawkins, Dennett is an avowed atheist.[3] Another prominent proponent of atheism—Sam Harris—has a background both in neuroscience (Ph.D. from UCLA), as well as philosophy (B.A., Stanford). [4]

By the same token, there have also been a number of recent cases of scientists plumping for theism, particularly for Christianity. Russell Stannard is both a Christian and a physicist who has chaired the Department of Physics at Britain's Open University and written a number of influential and award-winning works explaining contemporary physics to the public. He's also written a number of books, including *The God Experiment* (2000) and *Doing away with God? Creation and New Cosmology* (1993), that explore what he believes science does and, importantly, does not show about the existence of God. He is not alone as a contemporary theist scientist. Francis Collins, former leader in the Human Genome Project and current Director of the National Institutes of Health, has made his journey from atheism to faith the subject of a best-selling book, *The Language of God: A Scientist Presents Evidence for Belief* (2006). In addition, another eminent evolutionary biologist—Roman Catholic Ken Miller—has spent equal time arguing against Intelligent Design as an approach to understanding biology *and* arguing against the notion that contemporary biolog-

ical theory should be taken as supporting atheism. He does this most prominent-
ly in his *Finding Darwin's God: A Scientist's Search for Common Ground Between God and Evolution* (2000).

With all this recent discussion of the implications of science for either atheism or theism, agnosticism seems to get short shrift. This is an odd state of affairs given the long history of agnosticism and science. The term "agnostic" was
coined by nineteenth-century British biologist, Thomas Henry Huxley, although
the sentiment named by it probably goes back much further. The list of prominent professed agnostic scientists includes Albert Einstein (as quoted in Calaprice), Stephen Jay Gould, Francis Crick, and—if one believes political com-
mentator, John Lofton—Milton Friedman.[5] And, on the list of eminent
evolutionary biologists, there's none more eminent than Charles Darwin, who
aligned himself with his "Bulldog" Huxley on this point.[6] So, despite a recent
quietude on the part of agnostic scientists, there is a rich history of this pairing
of science and agnosticism.

So, simply doing a roll call of scientists and polling them on their attitudes
about the existence of God isn't going to get us very far. Scientists are as diverse
in their opinions as the rest of us. To answer the question of whether there is an
appropriate scientific attitude towards the existence of God, I propose that we
look at the relationship between scientific reasoning and the positing of supernatural explanations.

To start things off, let's take a closer look at Huxley's motivations for coin-
ing his new term. As he presents it, the idea first came to him when he was a
young aspiring scientist attending a British organization known as the Metaphysical Society in the 1870s. This was a pretty high-powered group of intellectuals interested in the critical assessment of religious and philosophical beliefs.
His anecdote is worth quoting at length. As Huxley recounts, he was provoked
by:

> [The members of that Society who] were quite sure that they had attained a cer-
> tain "gnosis"—had more or less successfully solved the problem of existence;
> while I was quite sure I had not, and had a pretty strong conviction that the
> problem was insoluble.

He continues:

> So I took thought, and invented what I conceived to be the appropriate title of
> "agnostic". It came into my head as suggestively antithetic to the "gnostic" of
> Church history, who professed to know so much about the very things of which
> I was ignorant; and I took the earliest opportunity of parading it at our Society,
> to show that I, too, had a tail, like the other foxes. To my great satisfaction, the
> term took.

He goes on to explain what he takes the position to hold:

> [A]gnostics, they have no creed; and, by the nature of the case, cannot have
> any. Agnosticism, in fact, is not a creed, but a method, the essence of which lies
> in the rigorous application of a single principle. That principle is of great antiq-

uity; it is as old as Socrates; as old as the writer who said, "Try all things, hold fast by that which is good"; it is the foundation of the Reformation, which simply illustrated the axiom that every man should be able to give a reason for the faith that is in him, it is the great principle of Descartes; it is the fundamental axiom of modern science. Positively the principle may be expressed: In matters of the intellect, follow your reason as far as it will take you, without regard to any other consideration. And negatively: In matters of the intellect, do not pretend that conclusions are certain which are not demonstrated or demonstrable. That I take to be the agnostic faith, which if a man keep whole and undefiled, he shall not be ashamed to look the universe in the face, whatever the future may have in store for him.[7]

What this long passage makes clear is that, in proposing agnosticism, Huxley took himself to be encapsulating a view of appropriate reasoning itself, whether scientific or otherwise. But, in calling it the "fundamental axiom of modern science," he also makes it explicit that he believes agnosticism is the basis for science.

# Gould's Magisteria

In the essay where he refers to himself as a "Jewish agnostic," Stephen Jay Gould also introduces a concept of the relationship between science and religion he calls "NOMA" (for "non-overlapping magisteria"). A *magisterium* refers to the teaching authority of the Catholic Church. As Gould describes it, "magisterium,"

> is a word derived not from any concept of majesty or awe but from the different notion of teaching, for *magister* is Latin for "teacher." We may, I think, adopt this word and concept to express . . . the principled resolution of supposed "conflict" or "warfare" between science and religion. No such conflict should exist because each subject has a legitimate magisterium, or domain of teaching authority—and these magisteria do not overlap.[8]

The idea here is that science and religion deal with two different areas of intellectual life—science with phenomena in the natural world and religion with issues of the meaning of life, morality, etc.—and, hence, ought not to conflict. They do not operate within the same jurisdictions. For Gould, NOMA provides grounds for dealing with the apparent conflict in contemporary culture between evolutionary biologists and proponents of various forms of religious creationism (Scientific Creationism, Intelligent Design, and so on). For Gould (and the Roman Catholic Church from which he lifts this concept) there is no conflict between religion and science here because the matter of how biological human bodies came into existence in the form that they have is part of the magisterium of science, whereas questions about the morality of human behavior and our salvation are within the religious magisterium. Further, these magisteria are *non-overlapping*. Just as theologians have no business making pronouncements about how long ago the earth came into existence, scientists have no standing to make pronouncements about the existence of God. Hence, Gould's agnosticism.

Not surprisingly, many of the atheistic scientists discussed earlier do not care for Gould's NOMA. Religious claims do not simply get a pass simply because they are of a different sort from scientific claims. Or perhaps only *some* of these claims fall into a different domain. One reading of NOMA is that it tracks the fact/value distinction. That is to say, science's domain is the domain of *facts;* science gets to say what "is" the case. On the other hand, religion's domain are *values*; religion gets to say what "ought to be" the case. On this reading of NOMA, moral issues of right and wrong fall under the magisterium of religion, whereas issues of what are the facts of the matter are under the magisterium of science. Science is about knowledge; religion wisdom.

While this is a coherent reading of NOMA, there are a number of difficulties. First, it is not at all clear that would-be scientific atheists will buy the distinction here.[9] The idea that science *qua* science is fact replete and devoid of values is far from obvious, as many in late twentieth-century philosophy of science (such as Longino) have been at pains to show. But, more problematic is the realization that in the case at hand, this reading does the opponent of atheistic science no good once it is pointed out that claims of God's existence is (no surprise) *an existence claim.* That is to say, to claim that God exists (or not) is to make a factual, not a value, claim. The claim that God's existence is a Good Thing would be to make a value claim, but the bare claim of God's existence is to make a claim about the factual constitution of the universe. Indeed, it is this reading of NOMA and supernatural existence claims that underlies the assertions by Dawkins and his colleagues that science has standing to rule on the existence of God and that the evidence is in.

## Hanson's argument against agnosticism

The line of argument that scientific reasoning leads one to atheism, despite its recent effluence, has older roots. I would like to focus on one particular argument that comes from the philosophy science. Norwood Russell Hanson wrote on the credibility of metaphysical claims of God's existence. [10] He articulates an argument that he calls the "Agnostic's dilemma"—that seeks to show that agnosticism with respect to God's existence is not, upon close consideration, a logically stable position.[11] In essence, he argues that the very evidence (or lack thereof) that is often taken to support agnosticism is better taken as supporting atheism.

On the one hand, the agnostic agrees with the atheist that there isn't sufficient evidence to reasonably support the claim that God exists. All parties agree: there *could* be such evidence—Hanson presents an amusing thought experiment in which the clouds part and all the world's inhabitants hear an apparition of God thunderously dress him down: "I've had quite enough of your too-clever logic-chopping . . . in matters of theology. Be assured, N. R. Hanson, that I do most certainly exist." That is just to say the claim of God's existence cannot be settled *a priori*; one has to consider the factual evidence.[12]

On the other hand, the agnostic agrees with the theist that there isn't sufficient evidence to reasonably support the claim that God doesn't exist. Hanson rehearses the oft-heard retort to atheists that they cannot prove that God *doesn't* exist. Such a universal negative claim cannot be established. That is just the nature of such universal negative claims.

At this point, though, Hanson calls "foul," on the grounds that the agnostic is illegitimately changing the rules of the game when switching from the analysis of the atheist argument to the analysis of the theist one, the resolution of which puts the agnostic in a dilemma:

> He [the agnostic] begins by assessing 'God exists' as if he were a fact-gatherer. He ends by appraising the claim's denial not as a fact-gatherer, but as a pure logician. But consistency demands that he either be a fact-gatherer with both the claim and its denial, or else play logician with both. If he would do the former, then he must grant that there is factual reason for denying that God exists—namely that the evidence which purports to favor his existence is *just not good enough*. If he would play the latter game, however—if he would make logical mileage out of 'It is not the case that God exists' by arguing that it can *never* be established—then he should treat 'God exists' in precisely the same way. He must say not only that the present evidence is not good enough fully to establish the claim, but that it never could be good enough.[13]

Of course, you'll note Hanson goes beyond simply castigating the agnostic for the inconsistency of the position; he also presents what he thinks is the correct way out: atheism: "When there is no good reason for thinking a claim to be true, *that* in itself is good reason for think the claim to be false!"[14] He notes that the agnostic's policy of demanding formal proof of the case of non-existence claims leads to absurdity: "There is no proof . . . that a blue Brontosaurus does not exist in Brazil. There is no proof that the Loch Ness monster does not exist."[15] And so on, for any number of wild possibilities. But, extrapolating the agnostic's stated policy for analyzing the claim for the non-existence of God would seem to force us to the absurd conclusion that we ought to remain agnostic about the existence of the Loch Ness monster, blue Brazilian Brontosaurs, etc. Hanson points out that it is much more reasonable in such cases to be atheists with respect to these beasts:

> 'How can you be so sure' comes the retort to which the response must be: I *am* so sure because
>
> 1. people have looked, and they have not [been] found, and
>
> 2. there is no good reason for supposing that there are still good reasons (circumstantial evidence independent of looking and not finding) for supposing that such things do exist. We infer beyond appearances to the existence of gravity, the positron and life on Mars. But what appearances require us to infer beyond them to monsters and unicorns? Indeed, what possible reason could *you*, dear retorter, have for supposing there to be the slightest chance that such creatures do exist? Science now possesses the best factual grounds for denying precisely this.[16]

Such is Hanson's scientific and overwhelmingly common sense analysis of where the agnostic and the theist both go wrong (although the agnostic suffers the further charge of being logically confused in addition to being simply wrong on the basis of the facts). Notice that Hanson's discussion flies directly in the face of Huxley's own approach. Hanson's atheism stems not so much from science *per se* as it does from the general principles of reasoning. But where Huxley sees a call for agnosticism, Hanson sees the same situation as calling for atheism.

# The Agnostic Scientist

Let's recap: The question at hand is whether there is a particular *scientific* attitude towards the existence of God. Contemporary discussion is full of scientific atheists and scientific theists but, while currently underrepresented, scientific agnosticism has a long and rich history. One response to all of this discussion is Gould's principle of NOMA that proposes that the relationship between science and religion is one of *non-overlapping* jurisdictions, so attitudes on religious questions are independent of those on scientific ones. Hanson provides a clear counterargument to this view. In arguing against agnosticism, he proposes that it is reasonable to conclude, based on current evidence, that God does not exist (although he can imagine that the facts might have been different, in which case the conclusion that we ought to draw would have similarly differed). Further, he argues that agnosticism as a philosophical position is unstable. If agnostics were consistent in their reasoning, they would simply be atheists about God.

Against arguments such as Hanson's what's the agnostic scientist to do? Is there a principled response on behalf of scientific agnosticism that can create a logically stable position against the worries of Hanson? I believe there is.

The first step in seeing a way forward here is to note that there is at least one relevant difference between God, on the one hand, and Hanson's blue Brazilian Brontosaurus and the Loch Ness monster, on the other. First, these mythical beasts are not generally thought of as agents who could be expected to be aware of our looking for them and having the resources to evade detection. Of a blue Brontosaurus in South America, it makes sense to say, as Hanson does, that "people have looked, and they have not been found," and that therefore we have grounds for saying they do not, in fact, exist. That is because we have no reason to believe the Brontosaur could successfully *hide*. But if God (or another supernatural agent) were omniscient—indeed if they were merely more intelligent and resourceful than we are—our looking for and not finding them should not lead so quickly to Hanson's conclusion.

The implications of God as a powerful, knowledgeable agent is one that I explore in Keeley.[17] There I point to the structural similarities between positing a role for God in the universe with the positing the role of powerful agents in certain popular conspiracy theories. In conspiracy theories and theology, but unlike in everyday natural science, I argue that one must take some account of the likelihood that the object of investigation has both the ability and the desire to avoid detection. Powerful agents (and groups of agents) have the ability to cover their tracks or otherwise throw pursuers off their trail. Disinformation is par for the course in such cases.

(I should also note here that this expected activity on the part of the object of investigation is one of the differences between the social and the natural sciences. Natural scientists do not typically have to spend as much time worrying about what their objects of investigation know about the studies they are carrying out. They don't typically need to worry about their rats overhearing their discussion of research protocols or be concerned that their Large Hadron Collider is in collusion with the hadrons themselves to deceive us about the nature of subatomic particles. Social scientists, and those natural scientists who work with humans, must pay much closer attention to these details. For this reason, scientific methods drawn from the natural sciences and applied to the study of motivated and resourceful agents should be *prima facie* suspect.)

I will not say more here about this idea of God as the ultimate conspiracy theory and instead turn my attention to a difficulty that Hanson raises that is not addressed by that line of argument. God may differ from the blue Brontosaurus because Brontosaurus's are not supposedly aware of our looking for them, but this is not the case with unicorns and apparently *magical* beings such as the gnome at the end of the garden (who are smart and powerful enough to disappear whenever any attempt is made to detect their existence). Surely science has grounds for denying the existence of magical or other supernaturally powerful beings?

Perhaps, but first a word about what constitutes *magic*. Here, I am going to conflate "magic" and "supernatural" as that which is beyond the known laws of nature. Why do I hedge by saying the *known* laws? Because we are always reasoning from our current standpoint and it has often been observed that what seemed to be supernatural (not possible given the laws of nature) in one era have been shown to be shortsighted in another.[18] Newton's gravity was action at a distance. Contemporary quantum cosmology violates *ex nihil nihilo fit* ("out of nothing nothing comes"). Master of speculative fiction, Arthur C. Clarke captured the notion well when he famously said in what is sometimes called *Clarke's Third Law*: "Any sufficiently advanced technology is indistinguishable from magic."[19] And what goes for technology also goes for science. The posits of any sufficient advanced science is indistinguishable from magic, as well.

If Clarke is correct then the would-be atheist scientist is faced with a conundrum of his own: To deny the existence, *a priori*, of the supernatural is to presume that we *know what* constitutes the genuinely supernatural (as opposed to what is merely beyond our current understanding of the natural world). But the history of science is littered with now dubious claims about not only what is currently unknown, but what is unknowable.[20] The problem here is that while we may have a pretty firm grasp on what is *unknown*, we have much less confidence with respect to that which is *unknowable*.

Part of the difficulty in knowing what currently unknown things will one day be understood and which will remain (forever?) unknown is that we do not currently know what epistemic access scientists in the future will have. Could ancient astronomer Ptolemy have predicted that we would one day have devices like the Hubble Space Telescope to peer at the stars from outside of earth's atmosphere (especially given that he lived 1500 years prior to Evangelista Torricelli's proposal that we live at the bottom of a sea of air that surrounds the entire planet)? Could eighteenth century phrenologist Franz Joseph Gall have guessed that within 200 years, we would have a technology-functional magnetic resonance imaging—that would allow scientists to tell non-invasively what a person was thinking?[21] Such knowledge about the nature of the solar system or the contents of people's minds might well have seemed genuinely unknowable to such scientists in their day.

This, then, is the ultimate source of scientific agnosticism: a humility about our current state of knowledge, not just about the facts but also the limits of knowledge itself. The agnostic scientist, such as Huxley, wishes to impress on us that we do not what the future of science holds. As a result, science, by its very nature, is open-ended. As scientists, we never fully say "case closed" because we never know how our entire view of what science is and what it is capable of might undergo revolutions at some point after our time.

Hanson might ask at this point—as he does in the passages quoted above—"what possible reason could *you*, Professor Keeley, have for supposing there to be the slightest chance that such creatures as unicorns and garden gnomes do exist?" My answer is that I do not have the slightest reason for asserting the existence of such creatures. But neither are there great numbers of otherwise credible people in the world who assert the existence of such creatures. And neither are there books published by respected scientists arguing for their existence. The case of God is different from unicorns and garden gnomes on this point, just as the case of God is different from Hanson's blue Brazilian Brontosaurus case, discussed earlier.

Can I lay out, in detail, the conditions of test and proof that would settle this matter at some point in the future? Can I explain to scientists what tests they need to perform in order to determine the existence or non-existence of this particular entity? No. But at the same time, I wish to insist that we simply do not know now what will be knowable or unknowable in the future. If there were thousands of apparently sane people claiming that they had encountered unicorns, as well as respected scientists doing the same, then I might be an agnostic, rather than an atheist, about unicorns. But there are not, so I am not. On the other hand, in the case of God, there are a great number of individuals (not including myself) who claim to have encountered such a being in their own lives. With Huxley, this leads me to an agnostic conclusion about the existence of God and, further, I believe—again with Huxley—that this reasoning is of a piece with what scientific reasoning would calls for.

A teaching colleague of mine used to have automobile with the following bumper sticker on it: "Militant agnostic: I don't know and you don't either!" This is a humorous way of presenting the Huxley point here. Huxley namechecks Socrates in his defense of agnosticism. Socrates was similarly militant in holding both that he himself did not possess knowledge and, through his interrogations of his Athenian compatriots, that nobody else did either. A militant defense of science as a genuinely open-ended and ongoing investigation, full of surprise and counter-intuitive turns should lead the conscientious scientist to arrive at similarly agnostic conclusion with respect to the existence of God.

# Notes

Many thanks to both James Griffith and Peter Kung, whose comments on a previous paper of mine acted in part as the impetus for this one. (I doubt I have fully satisfied either of them with what I say here.) This chapter was written while I was fortunate enough to be a Visiting Fellow at the Sydney Centre for the Foundations of Science in 2010. Thanks also to our hard-working editor William Krieger for giving me a venue to express these ideas.

1. The phrase "x qua y" is a bit of philosophical jargon and means, roughly, "x as a y" or, more colloquially. "x when wearing the hat of y." A given scientist qua scientist might not have any particular views on who ought to win a sports match, but that same scientist qua lifelong resident of New Orleans and avid Saints fan might have a strong and understandable preference. Scientists are human beings and as such have many different beliefs, especially on topics that aren't particularly related to science. So, in asking this question, I am asking whether scientists in their role as scientists ought to have any particular view on the the existence of God.

2. Dawkins, *The God Delusion* (New York: Bantam Books, 2006).

3. In a discussion of a brush with death (as a result of a "dissection of the aorta") that is simultaneously serious and humorous, Dennett describes his reaction to learning that religious friends of his had prayed for his health: While he sincerely appreciates their concern on his behalf. . . . "I am not joking when I say that I have had to forgive my friends who said that they were praying for me. I have resisted the temptation to respond 'Thanks, I appreciate it, but did you also sacrifice a goat?' I feel about this the same way I would feel if one of them said 'I just paid a voodoo doctor to cast a spell for your health.'" (Dennett, *Delusion*, 2006).

4. See Harris, *The End of Faith: Religion, Terror, and the Future of Reason* (New York: W.W. Norton, 2004); Harris, *Letter to a Christian Nation* (New York: Random House, 2006).

5. See Calaprice ed., *The Expanded Quotable Einstein* (Princeton: Princeton University Press, 2000); Gould, "Nonoverlapping Magisteria." *Natural History* 106 (1997): 16-22; Crick, *What Mad Pursuit: A Personal View of Scientific Discovery* (New York: Harper Collins, 1988); Lofton, "An Exchange: My Correspondence With Milton Friedman About God, Economics, Evolution And 'Values'". *The American View* (Oct-Dec 2006).

6. According to Darwin biographers, Desmond & Moore, in their *Darwin: The Life of a Tormented Evolutionist*. See pgs. 636 & 657.

7. Huxley, *Collected Essays*. Thoemmes Continuum; Facsimile of 1893-4 edition, 2001 237–239.

8. Gould, "Nonoverlapping Magisteria," 17 (emphasis in the original).

9. One who likely does not is philosopher of science, Philip Kitcher. See, in particular, the final chapter of his 2007 *Living with Darwin: Evolution, Design, and the Future of Faith*. There he argues that contrary to the soothing words from those apologists who say that religious belief has nothing to fear from science, the development of science is indeed a threat to religious forms of thought.

10. I discuss the significant influence of Hanson's work on at least one important contemporary philosopher of cognitive science in my introductory chapter to *Paul Churchland* (Keeley 2005). I also cover some of this same ground concerning Hanson on agnosticism in Keeley (2007).

11. Two of the papers he wrote on this subject—"What I Don't Believe" and "The Agnostic's Dilemma"—were published near the time of Hanson's unexpected death in 1967. Of the two, the former essay is the more comprehensive. Both papers are collected in the posthumous collection that I cite here (Hanson, Toulmin et al. 1971).

12. Hanson, N. R., S. Toulmin, et al. *What I do not believe, and other essays.* (Dordrecht: D. Reidel Publishing Company, 1971), 305-6, 313-4.

13. Hanson, et al., *What I do not believe,* 326 (emphasis in the original).

14. Hanson, et al., *What I do not believe,* 323, (emphasis in the original).

15. Hanson, et al., *What I do not believe,* 309-10.

16. Hanson, et al., *What I do not believe,* 324.

17. Keeley, "God as the ultimate conspiracy theory." *Episteme: A Journal of Social Epistemology* 4, no. 2 (2007): 135-149.

18. The attentive reader may have noticed that in one of Hanson's quotations above, he blithely refers to the existence of life on Mars in the same breath as affirming the existence of gravity and positrons. I am not sure why he does this, but I suspect that in the late 1960s there may have been evidence—prior to the Viking missions—to lead Hanson to make this claim with some confidence; a claim that in 2010 we would deny. This episode just underlines the point here of how scientific views change over time.

19. Clarke, A.C., *Profiles of the Future: An inquiry into the limits of the possible* (New York: Harper & Row, 1973), 39.

20. Just one such interesting case is the great German physiologist Emil du Bois Reymond (1818-1896). In addition to such achievements as discovering the action potential in neurobiology and giving birth to the science of electrical physiology, in 1872 he presented a series of seven "World Riddles" in his work *Über die Grenzen des Naturerkennens* ("On the limits of our understanding of nature"). Several of these riddles, such as the origin of life and the nature of free will, Du Bois Reymond argued were in principle solvable by science and philosophy, but that they posed great challenges to us. However, he referred to some of these riddles as ignoramus et ignorabimus, meaning "we do not know and will not know," including the ultimate nature of matter and energy and the origin of conscious sensation. These were beyond the ken of science and one wonders what he might have made of the twentieth century revolutions in physics, from Einstein's interrelationship of matter and energy and string theories posit of nine—and ten—dimensional space. These ideas of the following century certainly were beyond the known laws of science of Du Bois Reymond's time.

21. See, e.g., Singer, "They know what you want." *New Scientist* 183, no. 2458 (2004): 36-37; Phillips, "Private thoughts, public property." *New Scientist* 183, no. 2458 (2004): 38-41.

# Chapter 9

# Participating in a Contentious Natural Resource Debate as a Scientist: For Better or For Worse

## Yuri Yamamoto

Science is considered important, if not essential, for the management of natural resources.[1] Most, if not all, recent debates about integrating science and natural resource policy[2] focus on how to integrate science into policy rather than whether or not science has any role in policy.[3] While the benefits of having scientists involved in a natural resource policy debate seem clear, the benefits for scientists to actively participate in such a debate are much less obvious. In fact, such benefits seem few and far between because natural resource policy issues tend to be "messy" and "wicked."[4]

Many natural resource scientists consider performing research in a contentious debate as a risky endeavor.[5] A scientist working in a contentious arena risks being viewed as an advocate rather than as an independent researcher by his or her peers. Contending parties will question the credibility of a scientist who is perceived to be on the opposite side of the debate.[6] Policy-relevant projects can consume a large amount of time, yet the resulting technical reports or policy recommendations generally do not enhance a scientist's academic career.[7] Policy-relevant research questions are not highly valued within academic disciplines and are more difficult to publish in peer-reviewed journals.[8] During the Albemarle-Pamlico Estuarine Study, Korfmacher[9] also observed an inverse relationship between scientists' academic reputations and their willingness to work on questions useful for policy, which suggests that engaging in policy debates might be detrimental to a scientist's career (or vice versa). Scientists are also concerned that nonscientists misrepresent or selectively use the scientists' data for political gain.[10] Considering all the problems, scientists' reservation about performing research in a contentious setting seems reasonable, especially if they have to actually interact with contending parties.

With all these potential pitfalls, what is it like for scientists to actually be engaged in a policy debate? I interviewed academic scientists who participated in the North Carolina Wood Chip Production Study (Chip Study) from 1998 to

2000. The Chip Study incorporated extensive public participation, most notably the Advisory Committee, in which intense interactions between citizen stakeholders and scientists occurred. At the time of interviews, from 2003 to 2004, Chip Study scientists considered the Chip Study to be successful and wanted to "do it again" in spite of many challenges. This is noteworthy because many academic scientists stay away from research related to a contentious debate and are uncomfortable about public involvement. Experiences of the Chip Study scientists may shed light on how to effectively involve scientists in a contentious natural resource debate. In this chapter, I focus on why Chip Study scientists participated in the Chip Study, differences between their ordinary academic research and the Chip Study, and the scientists' overall assessments of the Chip Study experience.

# The Case

In the late 1990s, a group of academic scientists performed the Chip Study under the close scrutiny of the public.[11] Around this time, environmentalists were raising concerns about increased forest clearcuts in the U.S. South and beginning to campaign against large-capacity chip mills (Luoma 1997). Community groups of citizens were also concerned about aesthetic, noise, and traffic problems associated with chip mills and increased timber harvesting (Warren 2003; McClary n.d.). Many in the timber and paper industry, on the other hand, claimed that the chip mill was an extension of existing industry practice and was helpful for landowners.

Many Southern states responded to the controversy and began to examine their chip mill policies. In North Carolina, the governor commissioned a comprehensive study of economic and ecologic impacts of wood chip production. The study was led by the Southern Center for Sustainable Forests, a collaborative research center recently established between two universities and the state government to address forest sustainability issues in the south.[12] Academic leaders of the Southern Center assumed the Chip Study leadership, selected the Chip Study research agendas based on comments from public-scoping sessions, and recruited a group of academic scientists from the member institutions.

Recognizing the contentious nature of the debate, these leaders decided to extensively involve the public in the Chip Study process. A group of natural resource conflict-resolution facilitators developed a three-tiered public participation process that included the Chip Study Advisory Committee, public education forums around the state, and public comments on drafts of white papers and summaries. The most intense public involvement occurred through the Advisory Committee, consisting of a group of citizen stakeholders and government representatives.[13]

In the Chip Study, unlike many policy processes where scientists advise policymakers, citizen stakeholders and government agency representatives advised the scientists (Table 1). Chip Study citizen advisors, government agency representatives, and scientists interacted through the Advisory Committee meetings (six five-hour meetings and a field trip within a span of two years), ad hoc meetings, three public education forums around the state, and other means (telephone, e-mail, letters). At the first Advisory Committee meeting in 1998, an intense credibility contest emerged because environmental and community groups had serious concerns about the credibility of forestry scientists.[14] After several months of intense interaction, Chip Study participants overcame the credibility contest, and Chip Study white papers and summaries were approved by the Advisory Committee in 2000.

The final Chip Study reports included an Executive Summary, an Integrated Research Project Summary,[15] and eleven white papers covering a breadth of topics related to the controversy:

- North Carolina's Forests. Trends from 1938 to 1990;[16]
- Trends in North Carolina Timber Products Outputs, and the Prevalence of Wood Chip Mills;[17]
- Potential Wood Chip Mill Harvest Area Impacts in North Carolina;[18]
- Forest Resource Trends and Projections for North Carolina;[19]
- Soil and Water Effects of Modern Forest Harvest Practices in North Carolina;[20]
- Trends in Forest Composition and Size Class Distribution: Implications for Wildlife Habitat;[21]
- The Effects of Satellite Chip Mills on Post-harvest Woody Debris;[22]
- Storm Water and Process Water Management at North Carolina Wood Chip Mills;[23]
- Nonindustrial Private Forests: an Analysis of Changes in Potential Returns as a Result of Shifts in Demand;[24]
- Regional Economic Analyses of the Forest Products and Tourism Sectors in North Carolina;[25]
- Effects of Wood Chip Mills on North Carolina's Aquatic Communities;[26]
- Social Impact Assessment: Social Impacts and Community Concerns.[27]

While the Chip Study had been commissioned by the state governor, there was no integration between the Chip Study and a policymaking process. Chip Study reports were presented to the NC Department of Environmental and Natural Resources for further policy consideration. The Department made a number of policy recommendations, but no new policy was developed as a result.

# Research Methods

Detailed research methods are found elsewhere.[28] Briefly, semi-structured interviews were transcribed and coded based on grounded theory[29] and according to the social process framework of the policy sciences.[30] Archived documents were used for triangulation. I formally interviewed nine academic scientists on the Chip Study Research Team with expertise in resource economics, ecology, policy, and sociology; twenty-one stakeholder citizens; and one facilitator.[31] I also interviewed one administrator. All quotes in this article come from academic scientists unless otherwise noted. In order to protect scientists' identities, only the interview month is shown for each quote.

# Results And Discussion

## Reasons for participation

Why did these scientists participate in the Chip Study in the first place? The interests of the Chip Study scientists that emerged from interviews can be classified into three broad categories: interests in studying problems in society, interests in public participation, and career/funding opportunities.

Some Chip Study scientists had been involved in other policy-relevant projects such as the climate change assessments, riparian buffer policy, or development projects overseas. Quotes below exemplify the Chip Study scientists' interests and commitments to studying problems in society:

> My overall research interests are focused on wildlife and landscapes and my research mission . . . is to provide good scientific information to people whose activities shape our landscape.[32]

> I was interested in how modern forest management in the southeast is coping with environmental activist groups who have very different objectives from the land managers and a very different perception of the forest.[33]

> It made me interested in that we were actually addressing some real practical issues. . . . I got a big thrill in the whole study because I felt like the results of the outcome of the study will help drive a policy.[34]

Consistent with these interests, Chip Study scientists believed that science had a role in policy:

> [S]cience plays a role when people's perceptions are out there to determine whether or not their perceptions are accurate . . . So I think science can help tell if forests are being over cut, and we're losing forest. Science can help us tell if we're losing forest of particular critical species and habitat. Science can help us tell if we're getting more mud in the water and problems like that. Science can help us tell what size impact a chip mill may have. So I think that some of those questions that we addressed [in the Chip Study] helped confirm or reject hypotheses that groups had about the situation—and then probably just as much science did neither.[35]

> [A] minimal amount of technical evaluation often goes into decisions like this. I think that reflects a particular problem we have with dealing with science in contemporary society. We don't know where science is, and it's handled in a very uneven way. From my point of view, that's too bad. Science has a lot to contribute. It can't contribute everything by any means, but it has a voice, and it ought to be able to be a part of issues like the chip mill issue.[36]

The Chip Study was the first time that most scientists worked closely with citizen stakeholders. While Chip Study scientists had concerns about working with citizen stakeholders, their interests in and commitments to public participation were prevalent in interviews:[37]

[W]atching how public processes occur [elsewhere] . . . clearly we needed an inclusive, open, transparent, and adaptive process, and that's what we strove for.[38]

I'm a believer in [public participation].[39]

[The Chip Study] . . . looked very interesting because there were a lot of stakeholders involved. . . . You don't usually see that many people sitting down at a table attempting to guide scientific research—very unusual process. I was very curious as to how that would work.[40]

Such commitments and interests were not common among other scientists engaged in policy work, according to some Chip Study scientists:

I've been in other studies where there was a component that was labeled stakeholder, but [our] objective was to minimize that component as much as possible. In other words, "we have to show for this grant that there was a stakeholder participation, but do we really have to have a meeting? What else can we do?"[41]

[M]any of [my colleagues elsewhere] agree in principle [about the importance of public participation] for sure but most were more reluctant than we were to open it up to the extent we did. . . . they said, "I'm glad you were doing it, [and] we were not."[42]

Career and funding opportunities included summer salaries, postdoctoral positions, and graduate student assistantships. Because funding for the Chip Study was very limited, full-time university professors performed Chip Study research in addition to their full-time responsibilities without additional compensation. Some scientists declined to participate in the Chip Study because of the limited funding.

One of the Chip Study research leaders was a department head and encouraged his faculty to participate in the Chip Study. For a number of scientists, this was an important consideration. Even though they knew that Chip Study white papers would not count toward academic career advancement in general, "the fact that [the department head] asked me to participate and I said 'yes' . . . meant that I was going to get credit. It wouldn't be outside-the-university credit, but it would be within-the-department credit."[43] Having a partial extension appointment and working in a land-grant institution may also have encouraged some scientists to participate in the Chip Study.

Many natural resource scientists are interested in being useful to society so long as their value-free image can remain intact.[44] Scientists balance their desire to be useful, or to be perceived useful, with the potential risks to their credibility when deciding whether or not to participate in a policy-relevant study. For many Chip Study scientists, their interests in public participation, career/funding opportunities, and/or support from the department head, in addition to their interests in being useful to society, outweighed possible credibility risks.

Many Chip Study scientists had underestimated the credibility risks. It is unclear if these scientists would have participated in the Chip Study had they anticipated an intense credibility contest. Scientists who plan to participate in a contentious debate should be made aware of the credibility risks even though this awareness may discourage some of them from participating in such a debate.

## Differences between academic research and the Chip Study

Chip Study scientists experienced many differences between the Chip Study research and their ordinary academic research (Table 2). These differences primarily stemmed from the fact that the Chip Study was relevant to a contentious policy debate. Because all Chip Study scientists were academic researchers, the tension between academic success and policy needs was particularly noticeable.

### *Working with unusual colleagues*

Research involving participants across disciplines or beyond academia is a typical feature of sustainability science, while most academic research is conducted within disciplinary boundaries.[45] Due to the complex nature of research agendas, the Chip Study was multidisciplinary (scientists from more than one discipline were involved), interdisciplinary (scientists from more than one discipline worked together), and transdisciplinary (laypersons were involved).

Some of the Chip Study scientists conducted their parts of research independently. Integration of their portion of the report occurred at the level of writing the Integrated Research Project Summary. Even though these scientists came from multidisciplinary departments, most Chip Study scientists had not routinely worked with colleagues across disciplines. Therefore, they enjoyed getting to know their colleagues during long van rides to public education forums and learning from each other's presentations in meetings.

Interdisciplinary research to connect economic and ecologic impacts of future timber harvest was new scholarship and presented a variety of challenges. In order to assess potential impacts of changing harvest patterns[46] on wildlife habitat, economists and ecologists needed to work together. The Chip Study was the first time these scientists carried out such collaboration, and this was a new area of research.

Because of limited resources and time, the questions had to be addressed using existing data and modeling rather than collecting primary data. Because these disciplines used different assumptions, methods, and scales, translations and adjustments were necessary.[47] For example, the economic model tracked changes in forest composition at the level of North Carolina's regions (i.e, Northern Coastal Plain, Southern Coastal Plain, Piedmont, and Mountains), while ecologists are more accustomed to working within a local scale.[48] The economic model's ten-year age classes of five forest types (planted pine, naturally regenerated pine, mixed pine-hardwood, upland hardwood, and lowland hardwood) did not provide site-specific information related to particular requirements of individual species (e.g., presence of water nearby, large snags, well-developed midstory). Because the economic model was appropriate for analyzing only privately owned land, separate growth projections for public forestland (i.e., public parks, forests, military bases) had to be performed to complement the model's projections. Although this was hard work, it was a compromise rather than a new contribution from the disciplinary point of view.

The Chip Study was transdisciplinary in the sense that citizen advisors and government agency representatives worked very closely with the Chip Study scientists. Even though scientists conducted research and were the sole authors of the reports, citizen advisors heavily contributed to the Chip Study, including bringing their own data and editing reports. As an expert community, the Advisory Committee helped focus the Chip Study to address policy-relevant questions. Such transdisciplinarity is unusual in ordinary academic research.

## Different research focus

While academic research strives to produce generalized knowledge to advance theories, research conducted in a policy context is expected to provide specific, local information relevant to the particular issue. Most obvious is that the political boundary (i.e., the state of North Carolina) defined the geographical boundary of the Chip Study. This was appropriate only from a state policy perspective because the extent of federal policy, forests (natural boundary), forestry activities (economic boundary), and the chip mill controversy (social boundary) did not end at the state borders.

In the Chip Study, stakeholders' comments at scoping meetings defined research agendas, and citizen advisors' comments were incorporated into the Chip Study in many respects. For example, based on Advisory Committee recommendations, urbanization of forestlands became an important aspect of the Chip Study, and additional funding to study the impacts on aquatic species was obtained. Forestry Community members challenged the assumption that chip mills drove demand for wood and pointed out that public forested lands should be included in the wildlife habitat analysis. Environmental Community members questioned the judgmental term "nuisance species," requested a study on non-market values of forests, and urged scientists to collect primary ecological data. Citizen advisors of both communities submitted a variety of data pertinent to the Chip Study. Numerous comments and questions from both communities were incorporated into Chip Study white papers and summaries.

While Chip Study scientists remembered sometimes being yelled at from both sides, they were generally appreciative of the Advisory Committee members as demonstrated by the following quotes:

> [T]here was a tension in the room always to make sure that [we] were balanced. That was good. That's exactly what an advisory committee is supposed to do.[49]

[B]oth of those [additions to the research agenda] were enhancements to the study, made it a better study than it would've been if we hadn't addressed these issues.[50]

The advisory group was just wonderful, and these folks were so generous with their time. . . . everybody just made tremendous efforts to come to grips with the information and . . . proactively and constructively influenced the study.[51]

Chip Study scientists also acknowledged in a subsequent journal article that:

Concerns among stakeholders over the potential implications of preliminary study findings often led to probing questioning of both methodology and conclusions at the various study meetings. These questions and comments were at times disconcerting, but without doubt they enhanced the rigor of the study, and probably contributed to improved public acceptance of the ultimate research conclusions.[52]

These findings agree with the literature that stakeholder involvement in a fact-finding process improves the quality of information with regard to policy-making. Stakeholders help scientists study relevant questions and provide lay, local knowledge and values.[53] Stakeholders are also known to examine research assumptions, data, and interpretations critically—essentially functioning as a community of experts.[54] Furthermore, public involvement and critical examination of values and research processes may enhance research objectivity altogether.[55]

## Limitations of time and resources

The Chip Study resources were modest, and the timeline was fixed ($250,000 for two years for ten Research Team scientists and their graduate students to study economic and ecological impacts of chip mills in North Carolina). For most subjects, collection of primary data was not possible due to these constraints. The lack of adequate data also contributed to uncertainties and credibility problems.[56] Compared to funding for the Chip Study, academic research funding, though very competitive, can be much more generous and long term. Furthermore, scientists can continue academic projects to refine or expand theories with new funding.

Both Chip Study citizens and scientists were concerned about the limitations,[57] and Chip Study scientists had to come up with creative ways to complement the lack of data. For example, the U.S. Forest Service's Forest Inventory and Analysis (FIA), the single most reliable data for forest inventory, was last updated in 1990 for the region. This was a concern for all parties because the number of chip mills had increased dramatically in the several years before the Chip Study, and thus, most chip mill impacts on harvest trends would not have been captured by the latest FIA. Because timber harvests were not reported in North Carolina, there was no easy way to update the timber harvest trend required by the timber supply model. In order to make realistic timber growth/harvest projection scenarios, scientists had to estimate the recent harvest trend from the production data.[58]

## Controversial subject

Because the chip mill issue was a controversial subject, it was highly relevant to society, and public exposure was very high. In the Chip Study, the values of stakeholders were explicit in the debate. The majority of citizen advisors, however, expected Chip Study scientists to carry out their work without the influence of values, leaving the scientists' credibility in a precarious position.[59]

Within an academic discipline, shared values and intersubjectivity function as standards for objectivity.[60] Because Chip Study citizens and scientists did not necessarily share these standards, citizens questioned the credibility of scientists whom they perceived to be on the other side of the debate.[61] Most citizens did not recognize values similar to their own and expected objective scientists to authenticate their positions. While Chip Study scientists said that they did not have opinions about chip mills or that they could perform good research in the presence of biases, citizens' concerns about scientists' independence were mostly based on differences in the philosophy of broader forest management or in financial conflicts of interest.[62]

Practitioners of Joint Fact Finding claim that contending parties should select scientists by consensus to establish the credibility of scientists and their work.[63] Some Chip Study participants also reflected that the credibility contest might have been less intense had the Advisory Committee selected the Research Team scientists.[64] Nevertheless, the credibility of scientists in a contentious debate is fragile and must be managed throughout their interaction with contending parties.[65]

## Policy recommendations

Whether or not scientists should make policy recommendations was never resolved in the Chip Study. Because the issue remained controversial, scientists were caught between the Environmental Community's expectation to recommend policy options to regulate industry and the Forestry Community's insistence to refrain from it. The following quotes demonstrate that Chip Study scientists developed opinions about the chip mill issues over time and struggled to find the appropriate boundary:

> [I]t was very obvious to me some of the policy decisions that could be made, and I did have an emotional feeling towards some of those. I don't think that the role of the scientists can be so simply characterized as to stay within your technical box. It's impossible.[66]

> [S]urely I have opinions, but really I think they were not overtly influencing what we did or what we found, nor indeed does anybody in any of the advocacy groups care much, right? So they're just my opinions. I'm not a power broker out there influencing the outcomes of these policies as an advocate. I'm affect-

ing them more as providing good information that groups can use in their advocacy statements.[67]

> [A]t the end of the day, I was sort of waiting for someone to say, "well, you know, so what do you think about all this?" and no one did. . . . [B]ecause of the potential concerns of the bias and also just because of what we saw as appropriately our job to trying to generate facts here . . . we tried really hard just [to] stay on the fence over the life of the study, but I think all of us, over the course of the study, came to have our own opinions about things.[68]

Natural resource scientists indeed develop opinions about policy related to their research as evidenced by the prevalence of value-laden language and policy recommendations—or "stealth" advocacy—in natural resource journal publications.[69] Such stealth advocacy statements may go unnoticed within a discipline in which many members share the same values. In a contentious debate, however, these statements are vulnerable to criticism from those who do not share the values.[70] Furthermore, parties involved in natural resource policymaking have different opinions as to what extent scientists should be involved in policy.[71] Because the boundary between research and policy is flexible, it needs to be clarified during each process.

## Dealing with uncertainties

Like most natural resource policy debates, the Chip Study was fraught with uncertainties due to the resource constraints and complexity of the issues.[72] Dealing with uncertainties in a public setting was challenging to scientists. For example, at one of public education forums, an ecologist presented a map of chip mills and locations at which the presence of species of potential conservation concern had been recorded. On the map, these species of concern were absent around chip mills. The contending parties immediately interpreted the map in two opposite ways: (1) according to the Environmental Community members, the chip mill decimated these species and (2) according to the Forestry Community members, there was no concern about the chip mill's impact because these species did not exist nearby. The data actually could not distinguish these two interpretations because detailed spatial and temporal information was lacking. Furthermore, the species data were not based on comprehensive surveys, so the species' absence on the map did not mean the actual absence of the species on the ground.

In an academic situation, scientists may find such a map intriguing for discussion, but in this public education forum, the inconclusive map only fueled the debate and was not viewed as useful. This particular analysis was eventually dropped from the Chip Study.

Toward the end of the Chip Study, the Advisory Committee and policymakers requested scientists to prepare a brief Executive Summary targeted to policymakers who would not have the time or expertise to read and understand the entire body of white papers or the Integrated Research Project Summary. Archival documents showed that scientists were encouraged to generate simpler, shorter summaries with simpler tables. Scientists, while understanding this to be necessary for policymakers, were uncomfortable because detailed descriptions of uncertainties, assumptions, limitations, or interpretations would be lost. This was a major departure from the ordinary reporting practices of the scientists.

To cope with this difficulty, the Chip Study scientists decided to categorize findings in the Executive Summary using Mahlman's certainty levels[73] to caution readers of the limitation of the findings without getting into detail.[74] Ascribing certainties to findings in lieu of detail is not routinely practiced in ordinary academic research publications. Some Chip Study scientists found this process interesting and useful.

## Matters regarding communication

Because the audience of the Chip study was policymakers and stakeholders, scientists were constantly challenged to provide information in a form understandable to non-experts. Technical terms and assumptions were regularly questioned and clarified. Sometimes, miscommunication about research methods and expected outcomes appeared to have confused both citizens and scientists.

In the social impact assessment research, for example, the Advisory Committee requested the scientist to interview certain individuals in their own communities. While this requirement made the data anecdotal and statistically invalid, the same citizens were disappointed by the lack of quantitative proof of their lived experiences in the final report. Citizens were also perplexed by the report in which problems of chip mills were characterized as "perceived" rather than real. Archival documents and interviews revealed frustrations of both parties due to such miscommunication.

## "Gray literature" and academic careers

The Chip Study white papers and summaries fall in the category of "gray literature" and do not help advance scientists' academic careers.[75] Because academic scientists are expected to publish articles in peer-reviewed journals, most Chip Study scientists tried to publish their results in academic journals in their respective disciplines. Due to resource constraints and limited scope, some results have been more difficult to publish than others. The history of North Carolina forestlands, for example, was very important for the Chip Study but not suitable for most academic journals. A manuscript based on the wildlife component had been rejected twice because it was not a new finding. This is a serious drawback that discourages scientists from participating in policy-relevant research. A different reward system to recognize policy-relevant literature or policy involvement may help more scientists to engage in policy-relevant research without fear of compromising their careers.[76]

Many Chip Study scientists nevertheless found ways to publish their Chip Study results in academic journals.[77] One of these articles is a summary of Chip Study results, including the scientists' assessment of public participation.[78]

As more scientists engage in policy work, it would be ideal to develop more venues where gray literature can be turned into academic articles. Because resource constraints are the reality of policy-relevant research, it would be useful to share the successes and failures of creative approaches within these constraints rather than dismissing them as compromises and weaknesses. New theories and methods may be generated from sharing such experiences. Scientists' own reflections on their experiences of public involvement are rare and particularly informative to other scientists.[79] The Chip Study scientists' summary article is also unique in this respect.[80]

## Satisfaction of scientists in retrospect

Overall, the scientists considered their Chip Study experiences as positive. They remembered that, over time, citizen advisors' questioning had become "less acerbic, and a little more thoughtful"[81] in Advisory Committee meetings. Chip Study scientists also noted that all parties appeared to embrace the Chip Study reports in the end. Some scientists eventually developed positive relationships with skeptical citizens who had originally questioned the scientists' credibility.[82]

All scientists stated that the Chip Study research helped their subsequent careers or research directions, including participating in more policy-oriented projects. In retrospect, many scientists said that the Chip Study was a good experience and that they would participate in a similar project again:

> I think the experience has helped to make me a better scientist in the end, even though when we were involved in that pressure cooker situation, it was a little bit disconcerting sometimes.[83]

> On a topic like [the Chip Study] that has immediate policy relevancy, I think the advisory committee interactions were successful, and I'd want to repeat that experience.[84]

> I would do it again. I think it's difficult but rewarding and I think it adds something to the science, in terms of the public's perception of the end product and as far as their understanding of the complexity of the issues. It's good.[85]

Resolution of the credibility contest and successful completion and approval of the Chip Study reports was no doubt essential for the scientists' satisfaction in the Chip Study.

## Role of the Chip Study

While Chip Study scientists were satisfied overall, a lingering question remained in many of the scientists' minds: What was the role of the Chip Study in the chip mill and forest management controversy? During interviews, Chip Study scientists sometimes wondered if anything changed because of the Chip study. These scientists were interested in the Chip Study because the study would inform a contentious debate, perhaps even resolve it. In the end, however, voluminous white papers and summaries did not lead to any new policy. Was the Chip Study worthwhile?

Some may speculate that there never was a policy intent on the part of policymakers and that this was instead a political strategy to cool off the controversy or divert public attention. Regardless of the actual intent of the policymakers, the Chip Study had this effect. Shortly before the Chip Study began, the permitting process for new chip mills was made more stringent, and no new chip mill permit has been approved since the policy change. Within a few years after the Chip Study, some chip mills were closed due to economic changes. Overall, further proliferation of chip mills has not occurred, and the intensity of the controversy has subsided.[86]

Some scholars may say that the Chip Study was bound to fail in affecting policy because a successful fact-finding process should have been an integral part of a decisionmaking process.[87] Yet others would say that no new policy is a policy decision. According to Forestry Community members, the Chip Study results proved that no new regulation was needed. In contrast, Environmental Community members concluded that the Chip Study warranted new policies to regulate the industry. These disparate opinions demonstrate that the Chip Study did not settle the chip mill controversy. Neither did it change the minds of contending parties with respect to chip mills. Some scientists came to understand that the Chip Study situation was very complex and could not be resolved easily one way or another.

Participants in a contentious natural resource debate—citizens and scientists alike—tend to assume that science will resolve most, if not all, natural resource controversies and fail to see the role of values that underlie the controversy.[88] According to this assumption, the Chip Study was another failure—it did not produce a clear answer in terms of what to do with chip mills or forest management policy in North Carolina.

This science-to-policy management strategy, however, has not worked effectively in resolving conflicts in the last century because, as the Chip Study scientists learned, natural and human environments are too complex to be understood with certainty.[89] There is a new paradigm of natural resource management that incorporates science as one of many valid perspectives to inform policymaking.[90] From this new perspective of science as a piece of a complex policy puzzle, the Chip Study was successful. Most notably, the process of public participation made the Chip Study reports useful to all parties.

## Making science useful for policy

The Chip Study's utility is fundamentally different from the utility claimed by scientists of their own work.[91] When scientists claim utility of their work, they implicitly value science above other forms of knowledge and fail to understand the values underlying their own knowledge.[92] This kind of useful science has failed to resolve natural resource controversies under the scientific management principle.[93]

The utility of the Chip Study reports, on the other hand, was constructed through a consensus-building process by contending parties based on their unique values and knowledge. Some Chip Study scientists intuitively understood this and acknowledged that the reports would not have been accepted by the contending parties had the scientists conducted the study in isolation. By developing a single reliable source of information, contending parties moved the debate into a political arena. Debates about natural resources management are based on different worldviews and appropriately belong to politics.[94]

# Lessons Learned

Understanding and documenting subjective experiences of scientists in a contentious natural resource debate may help natural resource scientists, policymakers, and interested publics in many ways: those who want to be involved in contentious natural resource policy processes learn what to expect; those who participate in the debate have opportunities to reflect on their own experiences and how they might act in the future; and managers and policymakers gain insights into how to effectively involve citizens and scientists. Furthermore, the scientists' voices inform the researchers who study science and the science-policy interface. What can we learn from the experiences of the Chip Study scientists?

It is clear that some scientists are interested in stepping out of their comfort zone and contributing their expertise to a controversial debate with extensive public participation. Even for these interested scientists, differences between academic science and policy-relevant studies were at times confusing and challenging. In the end, however, these scientists were satisfied with their Chip Study involvement. An administrator who played a significant role for the evolution of the Chip Study offered three lessons learned from his observation of Chip Study.[95] Below, I integrate these lessons with the experiences of the Chip Study scientists.

## Lesson one: academic scientists should participate in policy-relevant science

"[I]t really is critical that the academic community on a selective basis participates actively in major public policy issues. . . . [F]or those that do have an inclination to engage in policy, there should be a way to reward them while they step out of their science and then allow them to step back into their science."

Most scientists would agree with this notion, considering that scientists generally believe that science has a role to play in natural resource policymaking.[96] However, far fewer scientists appear to be interested in actively participating in such a process.

Because not every scientist is interested in this type of work, identifying and enabling interested scientists is an important task. A better reward system and training through graduate school or professional workshops can help interested scientists to pursue policy-relevant work. Experienced facilitators of science-policy work[97] not only are essential for facilitating the public involvement process but also can train scientists who want to contribute their expertise to natural resource policy.[98] Developing a multidisciplinary, interdisciplinary, and transdisciplinary collaborative research center may also be helpful for providing such training to scientists by sponsoring policy-relevant research and altogether making this work a legitimate academic exercise (e.g., Southern Center for Sustainable Forests, which hosted the Chip Study).

Through training, scientists should learn about (1) how science is constructed in society and in scientific communities, (2) how values play a role in science and controversial debates, (3) what expectations and concerns citizens have about science, (4) what the challenges are of working with diverse colleagues, and (5) how to work with citizens who may be critical of the scientists. Importantly, interested scientists should learn from the experiences of scientists and citizens who have collaborated in policy-relevant research. Scientists should also learn about assumptions and methodologies of other natural resource disciplines in preparation for interdisciplinary research.

## Lesson two: scientists should establish credibility with stakeholder citizens

"[O]nce you step into that glare of that public light, you better have the best people around to do the job . . . people with impeccable credentials, impeccable credibility, impeccable integrity." What makes scientists' credibility impeccable? In the Chip Study, citizens questioned the credibility and integrity of renowned scientists because they had different values from the citizens.[99] In other words, one cannot have universal credibility just because he or she is highly respected by many people in academia or elsewhere in society. Based on my analysis of how citizens constructed the credibility of scientists, I have concluded that, to contending citizens, credibility is relative and that scientists working in a contentious debate must build and maintain a temporary sense of credibility through personal interactions with contending parties.[100] Establishing the credibility of scientists was imperative to making the Chip Study useful and was crucial for scientists' overall satisfaction in the Chip Study.

## Lesson three: scientists and stakeholder citizens should communicate with each other

"[Y]ou've got to have in place a process to be sure that those [scientists] are addressing the right issues . . . that are concerns of people who participated in the public policy issues. Then, you have to have a way to communicate with . . . the various stakeholders ad nauseam. . . . It's impossible to communicate too much on those issues." In the Chip Study, facilitators organized a three-tiered approach to establish effective, two-way communication with the public. Citizen input ensured that the scientists were asking questions relevant to citizens' concerns. Scientists strove to keep the research process transparent. The extensive communication through public participation was essential for making the Chip Study acceptable and useful in the end.

Creating a new community of experts takes a great amount of commitment by all parties. It can be painful and frustrating—and rewarding. Personal relationships built in such communities can last for a long time, and the lessons learned will carry forward into the future. For a natural scientist committed to public involvement, participating in a contentious debate with extensive public participation is for better, not for worse.

# Table 1. Roles of Chip Study Advisory Committee[101]

*Advisory Committee (citizen advisors)*
- Suggest to researchers issues to be addressed in the study or assist in establishing priorities of selected issues.[102]
- Recommend methods to collect and analyze data.
- Provide early feedback on study procedures and findings.
- Provide suggestions, support, and assistance for general public meetings.

*Technical Consultants (government agencies representatives)*
- Provide information and raise issues that are believed to be important for the Study Advisory Committee to consider.

# Table 2. Academic and policy-relevant research[103]

|  | *Academic research* | *Chip Study research* |
|---|---|---|
| *Research characteristics* | Generalized | Specific and local |
| *Participants* | Disciplinary peers | Multidisciplinary, interdisciplinary, and transdisciplinary peers |
| *Expected outcome* | Theory | Policy relevant information |
| *Audience* | Disciplinary peers | Policymakers and stakeholders |
| *Influence of values* | Shared values[104] | Value-laden/value-free[105] |
| *Resources* | Competitive, amount varies, renewable | Fixed and modest |

# Notes

1. Vogel and Lowham, "Building Consensus for Constructive Action: A Study of Perspectives on Natural Resource Management." *Journal of Forestry* (2007): 20-26; Cash, et al., "Knowledge Systems for Sustainable Development." *Proceedings of the National Academy of Sciences of the United States of America* 100, no. 14 (2003): 8086-91; Kates et al., "Environment and Development—Sustainability Science." *Science* (2001): 641-42; Lubchenco, "Entering the Century of the Environment: A New Social Contract for Science." *Science* (1998): 491-97.

2. Some natural resource issues can also be classified as environmental issues. I did not specifically distinguish these two concepts in this article.

3. E.g., scientific management versus adaptive governance discussed in Brunner and Steelman, "Beyond scientific management." In *Adaptive governance: integrating science, policy, and decision making*, by R. Brunner, et al, 1-46. (New York: Columbia University Press. 2005).

4. McCool and Guthrie, "Mapping the Dimensions of Successful Public Participation in Messy Natural Resources Management Situations." *Society & Natural Resources* (2001): 309-23; "Complexity, Wickedness, and Public Forests." *Journal of Forestry* 84, no. 4 (1986): 20-23.

5. Steel et al., "The Role of Scientists in the Environmental Policy Process: A Case Study from the American West." *Environmental Science & Policy* (2004), 1-13; Kinchy and Kleinman, "Organizing Credibility: Discursive and Organizational Orthodoxy on the Borders of Ecology and Politics." *Social Studies of Science* (2003), 879; Lach et al., "Advocacy and Credibility of Ecological Scientists in Resource Decisionmaking: A Regional Study." *Bioscience* (2003), 174; Korfmacher, "Invisible Successes, Visible Failures: Paradoxes of Ecosystem Management in the Albemarle–Pamlico Estuarine Study." *Coastal Management* (1998), 295.

6. Yamamoto, "Values, objectivity and credibility of scientists in a contentious natural resource debate." *Public Understanding of Science*, pre-published July 22, 2010, DOI: 10.1177/0963662510371435.

7. McDowell, "Engaged Universities: Lessons from the Land-Grant Universities and Extension." *Annals of the American Academy of Political and Social Science* (2003): 31-50.

8. Korfmacher, "Science and Ecosystem Management in the Albemarle–Pamlico Estuarine Study." *Ocean & Coastal Management* (2002), 277-300.

9. Korfmacher, "Invisible Successes."

10. Young and Matthews, "Experts' Understanding of the Public: Knowledge Control in a Risk Controversy." *Public Understanding of Science* 16, no. 2 (2007): 123-44.

11. Yamamoto "Values"; Schaberg et al., "Economic and Ecological Impacts of Wood Chip Production in North Carolina: An Integrated Assessment and Subsequent Applications." *Forest Policy and Economics* (2005): 157-74; Southern Center for Sustainable Forests, *Integrated Research Project Summary. Volume I: Economic and Ecologic Impacts Associated with Wood Chip Production in North Carolina*, 2000a; Southern Center for Sustainable Forests, *Working Papers. Volume II: Economic and Ecologic Impacts Associated with Wood Chip Production in North Carolina*, 2000b; Manuel, "Do Wood Chip Mills Threaten the Sustainability of North Carolina Forests?" *North Carolina Insight* 18 (1999): 66-93.

12. "The Southern Center for Sustainable Forests was established in 1997 by a Memorandum of Understanding among North Carolina State University, Duke University, and the North Carolina Division of Forest Resources. The Center was established to provide leadership for research, education, and extension to promote economically and ecologically sustainable management of forests in the South. Leadership for the Center is shared among the three co directors from each organizing institution. The Center sponsors and performs projects that involve research, teaching, and outreach regarding the practice of sustainable forest management in the South." Http://scsf.env.duke.edu/ (accessed on August 25, 2010).

13. Chip Study citizen advisors can be classified into two groups: Forestry Community members, who are dependent on forestry activities (e.g., timber product industry, consulting foresters, forestland owners), and Environmental Community members, who are concerned about and interested in regulating chip mill and forestry activities to pro-

tect natural and human environments (e.g., environmental and community groups and nature-based tourism industry). For more detail, see Yamamoto, "Values."

14. Yamamoto, "Values."

15. Southern Center for Sustainable Forests, *Volume 1.*

16. Cubbage et al., "Working Paper No. 1: North Carolina's Forests. Trends from 1938 to 1990." In *Economic and ecologic impacts associated with wood chip production in North Carolina, Volume II,* Southern Center for Sustainable Forests, 2000.

17. Schaberg et al., "Working Paper No. 2. Trends in North Carolina Timber Products Outputs, and the Prevalence of Wood Chip Mills." In *Economic and ecologic impacts associated with wood chip production in North Carolina, Volume II,* Southern Center for Sustainable Forests, 2000.

18. Dodrill et al, "Working Paper No 3. Potential Wood Chip Mill Harvest Area Impacts in North Carolina." In *Economic and ecologic impacts associated with wood chip production in North Carolina, Volume II,* Southern Center for Sustainable Forests, 2000.

19. Abt et al., "Working Paper No. 4. Forest Resource Trends and Projections for North Carolina." In *Economic and ecologic impacts associated with wood chip production in North Carolina, Volume II,* Southern Center for Sustainable Forests, 2000.

20. Richter, "Working Paper No. 5. Soil and Water Effects of Modern Forest Harvest Practices in North Carolina." In *Economic and ecologic impacts associated with wood chip production in North Carolina, Volume II,* Southern Center for Sustainable Forests, 2000.

21. Hess et al., "Working Paper No. 6, Part I: Trends in Forest Composition and Size Class Distribution: Implications for Wildlife Habitat." In *Economic and ecologic impacts associated with wood chip production in North Carolina, Volume II,* Southern Center for Sustainable Forests, 2000.

22. Hess and Zimmerman, "Working Paper No. 6, Part II: The Effect of Satellite Chip Mills on Post-Harvest Woody Debris." In *Economic and ecologic impacts associated with wood chip production in North Carolina, Volume II,* Southern Center for Sustainable Forests, 2000.

23. Greco and Gregory, "Working Paper No. 7. Storm Water and Process Water Management at North Carolina Wood Chip Mills." In *Economic and ecologic impacts associated with wood chip production in North Carolina, Volume II,* Southern Center for Sustainable Forests, 2000.

24. Snider and Cubbage, "Working Paper No. 8. Nonindustrial Private Forests: an Analysis of Changes in Potential Returns as a Result of Shifts in Demand." In *Economic and ecologic impacts associated with wood chip production in North Carolina, Volume II,* Southern Center for Sustainable Forests, 2000.

25. Aruna and Cubbage, "Working Paper No. 9. Regional Economic Analyses of the Forest Products and Tourism Sectors in North Carolina." In *Economic and ecologic impacts associated with wood chip production in North Carolina, Volume II,* Southern Center for Sustainable Forests, 2000.

26. Schaberg, "Working Paper No. 11. Effects of Wood Chip Mills on North Carolina's Aquatic Communities." In *Economic and ecologic impacts associated with wood chip production in North Carolina, Volume II,* Southern Center for Sustainable Forests, 2000.

27. Warren, "Working Paper No. 10. "Social Impact Assessment: Social Impacts and Community Concerns." In *Economic and ecologic impacts associated with wood chip production in North Carolina, Volume II,* Southern Center for Sustainable Forests, 2000.

28. Yamamoto "Values."

29. Strauss and Corbin, *Basics of Qualitative Research: Techniques and Procedures for Developing Grounded Theory*, 2nd ed. (Thousand Oaks: Sage Publications, 1998).

30. Clark, *The Policy Process: A Practical Guide for Natural Resource Professionals* (New Haven: Yale University Press, 2002).

31. See Yamamoto, "Values" for more information about the informants.

32. Interviewed in November 2003.

33. Interviewed in March 2004.

34. Interviewed in May 2004.

35. Interviewed in February 2004.

36. Interviewed in March 2004.

37. See Yamamoto, "Values" for more quotes.

38. Interviewed in October 2003.

39. Interviewed in October 2003.

40. Interviewed in April 2004.

41. Interviewed in November 2003.

42. Interviewed in October 2003.

43. Interviewed in November 2003.

44. For example, Kinchy and Kleinman, "Organizing Credibility: Discursive and Organizational Orthodoxy on the Borders of Ecology and Politics." *Social Studies of Science* (2003): 869-96, reported boundary-drawing discourses of Ecological Society of America members.

45. Blackstock and Carter, "Operationalising Sustainability Science for a Sustainability Directive? Reflecting on Three Pilot Projects." *Geographical Journal* 173 (2007): 343-57; Luks and Siebenhüner, "Transdisciplinarity for Social Learning? The Contribution of the German Socio-Ecological Research Initiative to Sustainability Governance." *Ecological Economics* (2007): 418-26; Kates et al., "Environment and Development—Sustainability Science." *Science* (2001): 641-42.

46. Scientists and Advisory Committee broadened most of the Chip Study research to encompass the overall timber product industry and harvest patterns.

47. Schaberg et al., "Economic and Ecological Impacts of Wood Chip Production in North Carolina: An Integrated Assessment and Subsequent Applications." *Forest Policy and Economics* (2005): 157-74.

48. Some of the examples were taken from a Chip Study white paper (Hess et al. 2000).

49. Interviewed in November 2003.

50. Interviewed in October 2003.

51. Interviewed in October 2003.

52. Schaberg et al. "Economic and Ecological Impacts," 170.

53. E.g., Karl et al., "A Dialogue Not a Diatribe—Effective Integration of Science and Policy through Joint Fact Finding." *Environment* (2007): 20; Brunner and Steelman, "Beyond scientific management." In *Adaptive governance: integrating science, policy, and decision making,* by R. Brunner, et al, 1-46. (New York: Columbia University Press, 2005); Cash et al., "Knowledge Systems for Sustainable Development." *Proceedings of the National Academy of Sciences of the United States of America* 100, no. 14 (2003): 8086-91.

54. I.e., "extended peer community" in Funtowicz and Ravetz 1991 and "epistemic community" in Korfmacher 2002: 284. See Collins and Evans 2002 on "expertise."

55. I.e., "concordant objectivity" and "interactive objectivity" in Douglas, "The Irreducible Complexity of Objectivity;" "strong objectivity" in Harding, *Whose Science? Whose Knowledge? Thinking from Women's Lives.* (Ithaca: Cornell University Press, 1991).

56. Yamamoto, "Values."

57. Yamamoto, "Values."

58. Schaberg et al., "Economic and Ecological Impacts."

59. Yamamoto, "Values."

60. Wallington and Moore, "Ecology, Values, and Objectivity: Advancing the Debate." *Bioscience* (2005): 873-78; Douglas, "The Irreducible Complexity of Objectivity." *Synthese* (2004): 453-73.

61. Yamamoto, "Values."

62. Yamamoto, "Values."

63. Karl et al., "A Dialogue Not a Diatribe," 20; Beecher et al., "Risk Perception, Risk Communication, and Stakeholder Involvement for Biosolids Management and Research." *Journal of Environmental Quality* 34, no. 1 (2005): 122-28.

64. Yamamoto, "Values."

65. Yamamoto, "Values;" Corburn, *Street Science: Community Knowledge and Environmental Health Justice, Urban and Industrial Environments* (Cambridge: MIT Press, 2005).

66. Interviewed in March 2004.

67. Interviewed in February 2004.

68. Interviewed in October 2003.

69. Lackey, "Science, Scientists, and Policy Advocacy." *Conservation Biology* (2007): 12-17; Scott et al., "Policy Advocacy in Science: Prevalence, Perspectives, and Implications for Conservation Biologists." *Conservation Biology* (2007): 29-35.

70. Yamamoto, "Values."

71. Vogel and Lowham, "Building Consensus for Constructive Action: A Study of Perspectives on Natural Resource Management." *Journal of Forestry* (2007): 20-26; Steel et al., "The Role of Scientists in the Environmental Policy Process: A Case Study from the American West." *Environmental Science & Policy* (2004): 1-13.

72. Sarewitz, "How Science Makes Environmental Controversies Worse." *Environmental Science & Policy* (2004): 385-403; McCool and Guthrie, "Mapping the Dimensions of Successful Public Participation in Messy Natural Resources Management Situations." *Society & Natural Resources* (2001): 309-23; Allen and Gould, "Complexity, Wickedness, and Public Forests." *Journal of Forestry* 84, no. 4 (1986): 20-23.

73. Mahlman, "Uncertainties in Projections of Human-Caused Climate Warming." *Science* (1997): 1416-17.

74. Southern Center for Sustainable Forests, *Volume 1.*

75. McDowell, "Engaged Universities: Lessons from the Land-Grant Universities and Extension." *Annals of the American Academy of Political and Social Science* (2003): 31-50; Korfmacher, "Science and Ecosystem Management in the Albemarle–Pamlico Estuarine Study." *Ocean & Coastal Management* (2002): 277-300; Korfmacher, "Invisible Successes, Visible Failures: Paradoxes of Ecosystem Management in the Albemarle–Pamlico Estuarine Study." *Coastal Management* (1998): 191-211.

76. McDowell, "Engaged Universities."

77. Snider et al., "Economic Analyses of Wood Chip Mill Expansion in North Carolina: Implications for Nonindustrial Private Forest (NIPF) Management." *Southern Journal of Applied Forestry* 30, no. 2 (2006): 102-08; Potter et al., "Multiple-Scale Landscape Predictors of Benthic Macroinvertebrate Community Structure in North Carolina." *Landscape and Urban Planning* (2005): 77-90; Schaberg et al., "Economic and Ecological Impacts;" Cubbage "Costs of Forestry Best Management Practices: A Review." *Water, Air, and Soil Pollution: Focus* 4, no. 1 (2004): 131-42; Murthy et al., "An Economic Analysis of Forest Products and Nature-Based Tourism Sectors in North Carolina." *Southern Rural Sociology* 20, no. 1 (2004): 25-38; Potter et al., "A Watershed-Scale

Model for Predicting Nonpoint Pollution Risk in North Carolina." *Environmental Management* (2004): 62-74; Warren, "Public Interests in Private Property: Conflicts over Wood Chip Mills in North Carolina." *Southern Rural Sociology* 19, no. 2 (2003): 114-31; Dodrill et al., "Wood Chip Mill Harvest Volume and Area Impacts in North Carolina." *Forest Products Journal* (2002): 29-37; Hess et al., "Working Paper No. 6, Part II: The effect of satellite chip mills on post-harvest woody debris." In *Economic and ecologic impacts associated with wood chip production in North Carolina, Volume II*, Southern Center for Sustainable Forests, 2000.

78. Schaberg et al., "Economic and Ecological Impacts."

79. E.g., Voinov and Gaddis, "Lessons for Successful Participatory Watershed Modeling: A Perspective from Modeling Practitioners." *Ecological Modelling* (2008): 197-207; Rhoads et al., "Interaction between Scientists and Nonscientists in Community-Based Watershed Management: Emergence of the Concept of Stream Naturalization." *Environmental Management* 24, no. 3 (1999): 297-308.

80. Schaberg et al., "Economic and Ecological Impacts."

81. Interviewed in November 2003.

82. Yamamoto, in preparation.

83. Interviewed in December 2003.

84. Interviewed in November 2003.

85. Interviewed in April 2004.

86. This may change in the near future because of growing interests in using wood chips for energy production.

87. Karl et al., "A Dialogue Not a Diatribe," 20; Beecher et al., "Risk Perception, Risk Communication, and Stakeholder Involvement for Biosolids Management and Research." *Journal of Environmental Quality* 34, no. 1 (2005): 122-28.

88. Yamamoto, "Values;" Sarewitz, "How Science Makes Environmental Controversies Worse;" Steel et al., "The Role of Scientists;" Rhoads et al. 1999

89. Sarewitz, "How Science Makes Environmental Controversies Worse."

90. E.g., Karl et al., "A Dialogue Not a Diatribe," 20; Brunner and Steelman, "Beyond scientific management;" Cash et al., "Knowledge Systems for Sustainable Development."

91. E.g., Kinchy and Kleinman, "Organizing Credibility: Discursive and Organizational Orthodoxy on the Borders of Ecology and Politics." *Social Studies of Science* (2003): 869-96; Gieryn, *Cultural Boundaries of Science: Credibility on the Line* (Chicago: University of Chicago Press, 1999).

92. Sarewitz, "How Science Makes Environmental Controversies Worse;" Rhoads et al., "Interaction between Scientists and Nonscientists in Community-Based Watershed Management: Emergence of the Concept of Stream Naturalization." *Environmental Management* 24, no. 3 (1999): 297-308.

93. Brunner and Steelman, "Beyond scientific management;"

94. Brunner and Steelman, "Beyond scientific management;" Sarewitz, "How Science Makes Environmental Controversies Worse."

94. Yamamoto, "Values;" Sarewitz, "How Science Makes Environmental Controversies Worse;" Slovic, "Trust, Emotion, Sex, Politics, and Science: Surveying the Risk-Assessment Battlefield (Reprinted from Environment, Ethics, and Behavior, 1997, 277-313)." *Risk Analysis* (1999): 689-701.

95. Interviewed in May 2004.

96. E.g., Vogel and Lowham, "Building Consensus for Constructive Action;" Kates et al., "Environment and Development;" Lubchenco "Entering the Century of the Environment: A New Social Contract for Science." *Science* (1998): 491-97.

97. E.g., Joint Fact Finding in Karl et al., "A Dialogue Not a Diatribe," 20.

98. Facilitation for Chip Study public participation was provided by the Natural Resources Leadership Institute. The institute provides training programs for natural resources professionals (Addor et al., "Linking Theory to Practice: A Theory of Change Model of the Natural Resources Leadership Institute." *Conflict Resolution Quarterly* 23, no. 2 (2005): 203-23), of which I was a participant. Such training programs can be modified to specifically train academic scientists. The U.S. Geological Survey also provides Joint Fact Finding workshops for scientists, of which I was a participant.

99. Yamamoto, "Values."

100. Yamamoto, "Values."

101. From Chip Mill Study Advisory Committee October 1998 Meeting Summary.

102. Researchers are referred to as "scientists" in this chapter.

103. Research directly or indirectly tied to a policy process. The Chip Study was mandated by the state governor and funded by the NC Department of Environment and Natural Resources. While many Chip Study participants expected a policy outcome, the Chip Study was not directly tied to a decision making process

104. While science is value-laden, shared values and intersubjectivity serve as standards for objectivity within a scientific community (Wallington and More, "Ecology, Values, and Objectivity: Advancing the Debate." *Bioscience* (2005): 873-78; Douglass, "The Irreducible Complexity of Objectivity.").

105. Citizens' values were explicit. Citizens, however, had general expectation for value-free science.

# Chapter 10

# Conflicting Values in Post-publication Disputes: The Case of Transgenic DNA in Mexican Maize

## Lawrence Souder

> In general, let every student of nature take this as a rule, that whatever his mind
> seizes and dwells upon with peculiar satisfaction is to be held in suspicion.
> <div align="right">Francis Bacon, <i>Novum Organum,</i> Book I, LVIII</div>

Whatever Bacon meant by "peculiar satisfaction" seems subjective and contextual. Since Bacon, scientists and certain philosophers of science have tried to characterize the practice of science as value free, or at least free of the subjective. The study of nature is supposed to be guided by certain constitutive values like accuracy and consistency but hampered by subjective values like self-interest and self-promotion. If such values arise in research, advocates for a value-free science try to dissociate the subjective from the constitutive. Scientists and students of science have weighed in on this dichotomy. Hempel contrasts the "questions of moral valuation" with "the objective methods of empirical science."[1] More recent commentary has not been so categorical. Kuhn contends that scientists make theory choices based on scientific values of accuracy, consistency, scope, etc., but because these value terms are imprecise and, when applied concurrently, conflictual, scientists need to fall back on shared subjective factors. Lacey contends that scientific and social values touch but do not interpenetrate each other. Longino tries to dissolve the dichotomy between the scientific and the contextual by arguing that scientists implement scientific values socially through interactive, discursive practices. More radically constructivist views like Barnes argue that all science is essentially social and that the empirical is not sufficient for scientific consensus.

Although the regard for the boundary between scientific and social values has become increasingly blurry, the scientific community itself continues to assert that boundary particularly in the context of consensus formation. Since Bacon, they have used among other things the peer review process as a means of sifting the social from the constitutive, what is regarded now as part of the self-correcting nature of science. However, reviews of peer review, like most things, have been mixed. At best peer review is a screen for gross incompetence and fraud, not a perfect filter. More troubling is peer review's tendencies to be inefficient and to preserve the status quo of male, orthodox science. Instances of research fraud like the Darsee case suggest peer review is little more than window dressing and that the real filter is the selection process subsequent to peer review—the dialogue in the research community that includes letters to the editor, errata, and retractions.[2] Nevertheless, exposés of fraud notwithstanding, as Alan Gross remarked: "Doubts about peer review are like doubts about democracy. Both systems are far from perfect, but in the case of peer review no alternative to the process of rational consensus seems more likely to work in the interest of furthering knowledge."[3]

The peer review process is a cornerstone of the pragmatic side of science; it determines who gets published, who gets funded, and who gets promoted. Such an institution, like democracy, should be subject to reassessment and revision. How to do this? Since peer review operates on the basis of values clarification, an assessment of the process might best begin with questions of values. What values drive the peer review process? Are these values different from those that drive the practice of science? Atkinson, in fact, claims that in science "peer review is remarkably unscientific."

In this chapter I try to assess the values that drive the peer review process, especially that part of the process that adjudicates who gets to participate and what is permitted on the agenda. To answer these questions I take a rhetorical approach to analysis of the process. Specifically I use Longino's notion of the ideal epistemic community to examine the texts of the discursive interaction among authors, editors, and referees as it evolves after the publication of research reports. In particular, I try to ascertain what values underlie their remarks and how these values shape their attempts to set the agenda of the discussion and to limit the membership of the community. My focus is on cases where the peer review process does not always terminate with publication; often post-publication errata protract the peer review process. In a recent (Quist) case, in fact, post-publication letters to the editor forced the journal to issue an erratum invalidating the original published research article. When the original researchers contested the erratum, the rational consensus process seemed to deteriorate.[4] In particular the Quist case seems to bring into high relief the conflict between scientific and social values. I conclude that social values become integral to the scientific discussion but ultimately have little effect on the judgment of the research and more problematically, they drive the debates over research validity into protracted and useless disputes over integrity.

Some quantitative studies report on the outcomes of peer review and its aftermath. Pfeifer and Thomsen found that research reports for which errata or retractions had been published continued to be cited in their original form by other researchers; from this result they suggest that the safeguards for correcting for bias in the peer review process are not working very well.[5] Much has been written about improving the process: Wilson offers a comprehensive set of guidelines for authors and referees,[6] and Stamps adapts a legal practice to peer review and offers a "dialectical scientific brief" as a way to streamline the process.[7]

The Quist case seems to test the putative constitutive values of science. One of those values which, as Charlton notes, makes science possible in the first place is its efforts to "filter-out personal partiality."[8] Many have already pointed out partiality in the practice of science. Carlton notes how commercial interests influenced clinical practice and health policy. Mallard shows how epistemological conflicts of interest compromise the impartiality of reviewers of funding applications.[9] Harrison describes his experience with partiality after his decision to publish Bjorn Lomborg's *The Skeptical Environmentalist*.[10]

# Framework for Analysis

Atkinson offers an account of "peer review culture" filled with anecdote, but the literature on peer review contains few extended case studies. A notable exception is Gross which treats peer review as an a Habermasian ideal speech situation, where each speaker: (1) has equal opportunity to initiate speech, (2) is permitted to reveal feelings and emotions, (3) may use any kind of speech act, (4) has equal power, and (5) is allowed to raise the level of discussion.[11] Submitting a research paper to peer review is a speech act (viz., a request) but as Gross finds, the peer review process is anything but ideal because a real-world author (1) has limited opportunities to initiate dialogue after the initial request, (2) must always be civil, unlike referees, (3) can not issue commands, (4) has little power over editors, and (5) has little recourse to appeal. In short, Gross's analysis highlights the power inequities in peer review process.

Gross's use of Habermas's ideal speech situation reveals important aspects of the speech act nature of the submission of research papers for publication and of the character of the agent of that act, but it tends to ignore the community's responses to the act and its actor. In short, this approach ignores the interactive nature of peer review. Longino notes that the peer review process illustrates the interactive or social character of science—particularly on those occasions when the peer review process falls short of its ideal.[12] Social values become salient, for example, when referees allow an author's institutional affiliation to sway their judgment of the author's submission. At this point, says Longino, the social functions of the back-up system to peer review-replication, subsequent citation, modification, or even retraction of the original published research—come into play to correct for any biases that may have compromised the initial peer review process.

It's this post-publication phase of peer review that requires a wider net to analyze than the model of the ideal speech situation provides. Longino offers such a net in her notion of the ideal epistemic community.[13] This construct can account for the way scientists reach rational consensus: they must submit their work to the challenges and criticism of their peers. This process is bound by conditions of effective or transformative criticism:

1. Standard venue for presentation and submission for criticism
2. Willingness to embrace and act on criticism by community of scientists
3. Public standards for evaluating scientific practice
4. Tempered equality that derives from (a) the community's many potentially dissenting points of view and (b) community standards for membership in the community.[14]

She explicates these criteria further by noting some impediments to ideal epistemic community:

1. Corporations' efforts to privatize science and commodify its results
2. Lack of closure that may result from open-ended dialogues
3. Attitude that values the production of knowledge over the criticism of knowledge claims
4. Oppositional tendencies toward too liberal standards or too limited equality.[15]

These impediments are, in a sense, anti-criteria because each one counters one of the four transformative conditions.

These criteria, especially that of tempered equality, seem more rhetorical than Gross's approach and thus more suitable for analyzing a discursive process where an advocate for new knowledge must appeal to values of the community and the community's criticism of knowledge claims must appeal to the concerns of the advocate. I use this rhetorical regard for an audience's values to identify and analyze the discursive interactions that arise from the post-publication peer review process.

# The Case of Transgenic DNA in Mexican Maize

My data are the published correspondence, author and editor responses, and editorial policies relevant to the dispute over the discovery of transgenic DNA in Mexican traditional maize.[16] This case seems to begin as a garden variety research paper: the researchers submit their report to *Nature*, and it is published after peer review as "Transgenic DNA introgressed into traditional maize landraces in Oaxaca, Mexico." Their claim is a straightforward proposition of fact: "Our results demonstrate that there is a high level of gene flow from industrially produced maize towards populations of progenitor landraces."[17] Their methods and results sections offer justification for the validity and interpretation of their data, and among their conclusions is the characteristically prudent standard appeal for further research, "Further study of the impact of the gene flow from commercial hybrids to traditional landraces in the centres of origin and diversity of crop plants needs to be carefully considered with respect to the future of sustainable food production."[18] Implicit in their argument seem to be the conventional scientific values of explanation and objectivity.

The first wave of criticism that Quist and Chapela provoke responds to similar values. Christou complains about the validity of their data: "the data presented in the published article are mere artifacts resulting from poor experimental design and practices."[19] He bases this charge by appealing first to the value of internal consistency: "the sequences obtained by PCR are rather strange,"[20] and "The inverse PCR results are problematic, internally inconsistent and not what is expected from cross pollination by commercial transgenic maize."[21] Christou then appeals to the value of external consistency by citing how the data does not comport with general understandings of the field: "This calls for a most unlikely recombination event."[22]

Kaplinsky criticizes Quist and Chapela on the same grounds: "[I]t is likely that the i-PCR sequences are all artifacts and not genuine transgenes."[23] Likewise Metz asserts: "we show here that their evidence for such introgression is based on the artefactual results of a flawed assay."[24] Nothing said so far in this scientific exchange seems untoward.

What makes the commentary of Kaplinsky and Metz unusual is that it was accompanied by a pronouncement by *Nature* editor Philip Campbell:

> In our 29 November issue, we published the paper "Transgenic DNA introgressed into traditional maize landraces in Oaxaca, Mexico" by David Quist and Ignacio Chapela. Subsequently, we received several criticisms of the paper, to which we obtained responses from the authors and consulted referees over the exchanges. In the meantime, the authors agreed to obtain further data, on a timetable agreed with us, that might prove beyond reasonable doubt that transgenes have indeed become integrated into the maize genome. The authors have now obtained some additional data, but there is disagreement between them and a referee as to whether these results significantly bolster their argument.
>
> In light of these discussions and the diverse advice received, *Nature* has concluded that the evidence available is not sufficient to justify the publication of the original paper. As the authors nevertheless wish to stand by the available evidence for their conclusions, we feel it best simply to make these circumstances clear, to publish the criticisms, the authors' response and new data, and to allow our readers to judge the science for themselves.[25]

Instead of letting the commentaries speak for themselves, *Nature* editors inserted their own position which included a shift from matters of data validity to those of reader autonomy and independence (though, they seem to both assert their view and invite the reader's at the same time).

In the same issue of *Nature*, Quist and Chapela respond to the criticisms of Kaplinsky and Metz by maintaining a focus on the original values—data validity and consistency. Their answer seems humble but firm. They concede on matters of validity: "We acknowledge our critics' assertion of the misidentification of sequences; [but stand firm on matters of consistency] . . . [our work] was confirmed by Mexican government studies . . . [and] Our findings are compatible with recent studies."[26] Other commentators weigh in on the issue of validity, but the end of this round of the debate is anti-climactic as summarized in the carping of Wager:

> The information provided falls short of proving the presence of transgenes, and clearly cannot support any claims of introgression of transgenes into Mexican land races. Nevertheless, it is probably inevitable that eventually engineered genes will be found in Mexican corn, as gene flow is a normal and natural phenomenon with maize. Such gene flow-by chance or intent—has given rise to a large amount of biodiversity, which is balanced as farmers select for specific characteristics that make each local maize variety unique. As is there is no reason to believe transgenic corn will pose a greater or lesser risk to native varieties than any of foreign malze varieties already brought to Mexico.[27]

In short the critics end up having it both ways: by impugning the empirical data but speculating about the probability of its eventual truth, they both acknowledge the presence of transgenes and doubt it.

The second wave of criticism of Quist and Chapela seems to depart from the first by shifting from issues of research validity to those of researcher integrity. At this point the debate includes ad hominem remarks directed at researcher, referees, and editors. These remarks include allusions to *Nature*'s published editorial policy on "competing financial interests." Overall, *Nature*, as a scientific publisher, is concerned with the value of impartiality, and so it institutes policies "[i]n the interests of transparency and to help readers to form their own judgments of possible bias."[28]

This impartiality depends on the value of an informed audience, and so *Nature* carefully and fully articulates a policy on conflicts of interest, which it defines as "those of a financial nature that, through their potential influence on behaviour or content or from perception of such potential influences, could undermine the objectivity, integrity or perceived value of a publication." Authors, *Nature* says, should disclose their "competing financial interests." Referees, *Nature* says more broadly, should disclose "any related interests, including financial interests." And editors, *Nature* says more broadly still, should disclose "any interests—financial or otherwise—that might influence, or be perceived to influence, their editorial practices." In this tripartite articulation the scope of the term *conflict of interest* progressively broadens to eventually encompass even the appearance of influence. However, after having so carefully justified and articulated these principles of disclosure, *Nature* seems to take them back, saying: "However, just as financial interests need not invalidate the conclusions of a paper, nor do they automatically disqualify a referee from evaluating it."[29]

This apparent equivocation over the value of impartiality not withstanding, Suarez begins the second round of the debate with a general invocation of a standard of scientific publication:

> Many journals include a forum for scientists to make technical comments on recent publications and for the original authors to respond, so that readers can evaluate the merit of reported scientific findings. This thorough evaluation ensures, to some extent, the journal's impartiality in publication decisions.[30]

From this moral high ground, Suarez attacks *Nature*'s quasi-retraction of Quist and Chapela's original paper: "By taking sides in such an unambiguous manner, *Nature* risks losing its impartial and professional status."[31] Then Suarez seems to cross the line that divides the scientific from the social when he says: "This is particularly troubling when articles are related to economic or political interests."[32] The practice of science, it seems, is now not wholey dependent on the empirical.

In the same issue of *Nature*, Worthy moves from attacking research to attacking the critics: "The controversy surrounding Quist and Chapela's findings . . . is taking place within webs of political and financial influence that compromise the objectivity of their critics."[33] He specifically calls attention to the funding that the critics have received from a biotech firm, claiming: "None of the eight authors declares this funding as a competing financial interest in their published contributions."[34] Then, he recapitulates Suarez's criticism of the editors: "These partnerships seem to us to challenge *Nature's* ability to provide a neutral forum for scientific debates on agricultural biotechnology."[35] In a sense Suarez and especially Worthy argue for the value of impartiality because they claim its violation has corrupted the critical review of research validity.

In his riposte Metz attempts to dissociate financial interest from the practice of science: "[O]ur connections to industry are irrelevant to the scientific issues, and hence do not warrant disclosure . . . [and] Our concern was exclusively over the quality of the scientific data and conclusions, which would have been the same whatever the motivation of the criticism."[36] Kaplinsky likewise tries to enforce the boundary between the scientific and the fiduciary, saying: "funding from TMRI has absolutely nothing to do with our criticisms;"[37] and "Even if we were in the pockets of industry, Quist and Chapela's published results would still be artefactual."[38] However, Kaplinsky reverses himself by invoking financial interest as relevant after all when he says, "It is a double standard to accuse us, but not Quist and Chapela, of a conflict of interest."[39]

*Nature's* editor claims the last word in this round with the closing appeal to the value of validity: "[the] principal conclusion is shown to be not necessarily false but unsustainable on the basis of the reported evidence."[40] But impartiality is also invoked: "The paper was not formally retracted by its authors or by *Nature*."[41] Finally, the editors dissociate themselves from all others who would smuggle social values into the practice of science when they proclaim: "The independence of our editorial decision-making from partisan anti- or pro-technology agendas and from commercial interests is paramount in our role as a journal."[42] *Nature* here is making a not-so-veiled reference to Kaplinsky and Metz's funding by the bio-tech industry and to Chapela's membership on the board of directors of Pesticide Action Network North America, a group that advocates alternatives to pesticides. What remains veiled, however, is the fact that, as Worthy points out, *Nature's* de facto retraction of Quist and Chapela's paper immediately preceded the sixth meeting of the Conference of the Parties to the Convention on Biological Diversity (COP 6) held in The Hague from April 7 to 19, 2002, a body that would develop binding global guidelines on genetic resources (UNEP). So it seems that what started out as a dispute over a claim for the existence of transgenic DNA in traditional crops turned into a debate over data validity, and, as *Nature's* capstone proclamation indicates, devolved into a political scuffle between alleged industry partisans and environmental activists over the integrity of science and scientists.

## Conclusion

What can we conclude about consensus formation from this analysis of some of the discursive record of the debate over transgenic DNA research? Longino posits tempered equality as an essential criterion for consensus formation in science. This criterion represents a difficult balance between widening the access to the discussion to ever more diverse points of view and maintaining standards for limiting the membership of the discursive community to a manageable and productive population. From this case study it seems that as the composition of the audience widened and opened the debate to vaguer claims, the debate moved from a scientific value (validity) to a social value (impartiality). The authors Quist and Chapela are affiliated with the Department of Environmental Science, Policy, and Management at the University of California. The peer review of this research occurred under the aegis of *Nature*, a multi-disciplinary scientific journal, whose coverage includes genetic research; but the first post-publication critique was from Christou, the editor of *Transgenic Research,* a journal limited to a subspecialty. Subsequent criticism came from Kaplinsky, Department of Plant and Microbial Biology at the University of California, and Metz, Department of Microbiology at the University of Washington.

The dispute over the validity of Quist and Chapela's research results seemed straightforward. In fact, the eventual consensus seemed to be that their results were more artifact than data. On the value of impartiality, however, the debate seemed more ambiguous. The claims for the relevance of impartiality to the debate were often equivocal. *Nature* required statements of disclosure from public participants in the debate but said later that financial interests don't necessarily invalidate one's research or one's critique of research. *Nature* firmly maintained their own impartiality in the debate but unambiguously offered a critique of Quist and Chapela's results that stopped short of an overt retraction. Claims for the irrelevance of impartiality to the debate seemed disingenuous when they were followed by counterattacks of partiality. Harrison put it more bluntly: "The dividing line between science and politics is possibly nowhere thinner than in the case of . . . *Nature*'s publication of the Quist and Chapela paper."[43]

All of the appeals to impartiality seemed in defense of claims for validity, but the issue of validity seemed to be decided on other grounds. Appeals to impartiality instead seemed to push the debate from *ad rem* to *ad hominem* arguments. The criterion of tempered equality seems like a fine ideal, but its goal of democratizing science makes the introgression of a social value like impartiality so easily a catalyst for protracted but jejune disputes. This conclusion reinforces the importance of the separation of social from constitutive values in the practice of science.

# Notes

The first draft of this chapter was presented at the 2003 NEH Summer Institute on Science and Values at the University of Pittsburgh.

1. Hempel, Carl. "Science and Human Values." Pp. 82 in *Aspects of Scientific Explanation*, New York: Free Press, 1965.
2. See Culliton, B.J. "Coping with fraud: The Darsee Case." *Science* 220 (1983): 31-35.
3. Gross, Alan, 2003, personal email to the author. Sunday, Jun 22, 2003.
4. Quist, David and Ignacio H. Chapela. "Quist and Chapela reply." *Nature* 416, no. 6881 (April 11, 2002): 601.
5. Pfeifer, Mark P. and Gwendolyn L. Snodgrass. "The continued use of retracted, invalid scientific literature." *JAMA, The Journal of the American Medical Association* 263, no. 10 (March 9, 1990): 1420.
6. Wilson, J.R. "Responsible Authorship and Peer Review." *Science and Engineering Ethics* 8 (2002): 155-174.
7. Stamps, A.E. "Using a Dialectical Scientific Brief in Peer Review." *Science and Engineering Ethics* 3 (1997): 85-98.
8. Charlton, B.G. "Conflicts of interest in medical science: peer usage, peer review and 'Col consultancy.'" *Medical Hypotheses* 63, no. 2 (2004): 182.
9. Mallard, G., Lamont, M., & Guetzkow, J. "Fairness as appropriateness: Negotiating epistemological differences in peer review." *Science Technology Human Values* 34, no. 5 (2009): 573-606
10. Harrison, C. "Peer review, politics and pluralism." *Environmental Science & Policy* 7, no. 5 (2004): 357-368.
11. Gross, Alan. "Peer Review and Scientific Knowledge." Pp. 129-143 in *The Rhetoric of Science*. Cambridge, Mass.: Harvard University Press, 1990.
12. See Longino, Helen. *Science as Social: values and objectivity in scientific inquiry*. Princeton: Princeton University Press, 1990.
13. Longino, Helen. *The Fate of Knowledge*. Princeton: Princeton University Press, 2002, 129-135.
14. Longino, *The Fate of Knowledge*, xx, 129-135.
15. Longino, *The Fate of Knowledge*, 129-135.
16. Quist, David and Chapela, Ignacio H. "Transgenic DNA introgressed into traditional maize landraces in Oaxaca, Mexico." *Nature* 414, no. 6863 (November 29, 2001): 541-543.
17. Quist, David and Chapela, Ignacio H. "Transgenic DNA," 542.
18. Quist, David and Chapela, Ignacio H. "Transgenic DNA," 542.
19. Christou, Paul. "No credible scientific evidence is presented to support claims that transgenic DNA was introgressed into traditional maize landraces in Oaxaca, Mexico." *Transgenic Research* 11 (2002): iii.
20. Christou, Paul. "No credible scientific evidence," v.
21. Christou, Paul. "No credible scientific evidence," v.
22. Christou, Paul. "No credible scientific evidence," iv.
23. Kaplinsky, Nick, et al. "Biodiversity: Maize transgene results in Mexico are artifacts." *Nature* 416, no. 6881 (April 11, 2002) 601.
24. Kaplinsky, Nick, et al. "Biodiversity: Maize transgene results in Mexico," 600.
25. Campbell, Philip. Editorial Note. *Nature* 416, no. 6881 (April 11, 2002) 600.
26. Quist, David and Ignacio H. Chapela. "Quist and Chapela reply." *Nature* 416, no. 6881 (April 11, 2002): 601.

27. Wager, Robert, Peter Lafayette, and Wane Parrott. "Letter received on Wednesday 14 August 2002." *EJB Electronic Journal of Biotechnology.* http://www.ejbio technology.info/content/vol5/issue2/letters/01/index.html (accessed July 10, 2010).

28. See *Nature* magazine's editorial policies, accessible online. http://www.nature.com/authors/editorial_policies/competing.html (accessed July 10, 2010).

29. See *Nature* magazine's editorial policies.

30. Suarez, Andrew V. "Conflicts around a study of Mexican crops." *Nature* 417 (2002): 897.

31. Suarez, "Conflicts around a study of Mexican crops," 897.

32. Suarez, "Conflicts around a study of Mexican crops," 897.

33. Worthy, Kenneth, Richared Strohman, and Paul Billings. "Conflicts around a study of Mexican crops [Correspondence]." *Nature* 417, no. 6892 (June 27, 2002): 897.

34. Worthy, Kenneth, Richared Strohman, and Paul Billings. "Conflicts around a study of Mexican crops," 897.

35 Worthy, Kenneth, Richared Strohman, and Paul Billings. "Conflicts around a study of Mexican crops," 897.

36. Metz, Matthew and Johannes Fütterer. "Biodiversity: Suspect evidence of transgenic contamination; Metz replies." *Nature* 416 (June 27, 2002): 897.

37. Kaplinsky, Nick. "Conflicts around a study of Mexican crops; Kaplinsky replies" *Nature* 416 (June 27, 2002): 898.

38. Kaplinsky, Nick. "Conflicts around a study of Mexican crops," 898.

39. Kaplinsky, Nick. "Conflicts around a study of Mexican crops," 898.

40. Kaplinsky, Nick. "Conflicts around a study of Mexican crops," 898 (editor's reply).

41. Kaplinsky, Nick. "Conflicts around a study of Mexican crops," 898 (editor's reply).

42. Kaplinsky, Nick. "Conflicts around a study of Mexican crops," 898 (editor's reply).

43. Harrison, C. "Peer review, politics and pluralism." *Environmental Science & Policy* 7, no. 5 (2004): 364.

# Chapter 11

# On Scientific Advocacy: Putting Values and Interests in Their Place

## Evelyn Brister

Scientific research informed many important public policy decisions of the late twentieth century. In the 1980s, depletion of the ozone layer threatened to increase rates of skin cancer, but the rate of depletion diminished following a 1989 ban on the production of chlorofluorocarbon compounds, primarily used in aerosol sprays. Likewise, acid rain made lakes in the northeastern part of the United States uninhabitable by fish, but the problem was mitigated when legislation required controls on industrial emissions of sulfur dioxide and nitrogen oxides. These are just two crises where scientists played a key role in identifying the causes of a significant social, health, agricultural, or ecological problem and in developing technological and regulatory solutions. We can expect public policy decisions to become even more dependent on science in the twenty-first century as we continue to confront problems such as global pandemics, climate change, the loss of biodiversity, and the release of toxic pollutants in air, water, and soil. These and other problems will open up opportunities for scientists and scientific professional organizations to play a larger role in promoting sound public policy.

But *how* should scientists interact with policy-makers? Does scientific expertise convey any relevant expertise in policy-making? Are the choices that individual scientists make to support one policy option or another *better* choices in virtue of their scientific expertise? Or do scientific credentials convey a form of epistemic authority which risks being *mis*used in order to secure illegitimate but influential moral authority?

Philosophers continue to debate the necessary extent and appropriate role for social and ethical values in the production of scientific knowledge. For most of the twentieth century, common wisdom among scientists, philosophers, and politicians held that only constrained, well-defined scientific advisory positions were legitimate, and as a result these were open to only a few scientists. Science was considered an autonomous enterprise which ought to be insulated from political action for the good of both science and policy.

But in recent decades, the general consensus among philosophers of science has shifted away from drawing a sharp distinction between descriptive science and prescriptive policy-making and toward a greater recognition that scientific work is informed by both facts *and* values. The resulting consensus is that science cannot and should not be conceived as value-free. Consequently, this debate over the value-freedom of science informs the debate over whether scientists can and should act as advocates. If the role of values in science, while not inevitable, is nonetheless potentially positive and desirable, then an inferential link can often be made between scientific findings and normative policy recommendations. If, in addition, values have an ineliminable role in scientific reasoning, then scientists may even have a positive duty to express their value-based commitments and to advocate on behalf of those commitments. This conclusion points to the legitimacy of advocacy roles while at the same time intensifying the risk that scientist advocates may misuse their authority.

In addition to the question of whether advocacy interferes with the process of scientific reasoning, there are also questions about how advocacy roles could affect the careers of scientists, public perceptions of scientific authority, and social interactions among scientists. Can one be a scientist and an advocate simultaneously? And if scientists begin to take on advocacy roles more frequently than they have in the past, how could such a shift in community practice affect the respect that is granted scientific communities? Does an increase in advocacy introduce problems that could adversely affect the functioning of scientific communities, and therefore the production of scientific knowledge? Such questions about the institutional structure of scientific communities and about how institutional practices support or undermine knowledge production have become key debates within the philosophy of science.[1]

Building on the work of other philosophers of science, I will largely assume that science is not value-free. Given this assumption, I will examine instead the proper place of values and advocacy in the scientific enterprise. These issues have received less attention from philosophers so far, even though they have long been a focal concern of scientists.[2] So, first, it is necessary to examine the debate over whether advocacy is *ever* legitimate for scientists.

## Objectivity comes in two flavors

There has long been a tacit prohibition on scientists' expressing political views while in their role as scientist. This prohibition stems from a number of concerns, all of them related to the possibility that political advocacy will undermine the objectivity of scientific research or the perceived objectivity of the scientific enterprise as a whole. However, objectivity has several meanings in relation to scientific inquiry.[3]

First, objectivity can refer to the idea that scientific knowledge represents the world as it (really) is: independent of human wishes, desires, and limitations. The problem, then, is that humans don't have access to the world as it (really) is except through their own experiences, and human experience is notoriously fallible. Too often, when we want to see something, we do see it. When we don't want to see something, we don't. When a phenomenon is brand new and surprising, we easily misperceive it or misinterpret its causes and results. There are hallucinations, sometimes even group hallucinations, and sometimes we discover new things under the sun which past experience cannot guide us to make good sense of. We might call objectivity in this first sense *metaphysical* objectivity, and its analysis is the dominion of philosophy.

Second, objectivity can also refer to a suite of methodologies that have been developed to decrease or eliminate bias and subjectivity. These are techniques that have become incorporated into scientific method through a process of trial and error, largely by scientists in search of improving scientific practice. These techniques assist scientists in making and evaluating observations in order to eliminate unreliable beliefs. Many are simple methodological rules, such as a preference for simple rather than complex explanations and ordinary rather than unusual causes. Objectivity is also supported by scientific norms, such as the requirement that results be reproducible. This kind of *methodological* objectivity is intended to certify that scientists have done all they can to rule out the possibility that their claims systematically misrepresent the world. Unlike metaphysical objectivity, the question of methodological objectivity focuses directly on scientific methods, practices, and conventions. For this reason, I will, likewise, focus on the methodological sense of objectivity here.

The methodological sense of objectivity is aimed at producing reliable claims—that is, claims that are less prone to require revision in the future. Effective methodologies for guaranteeing objectivity have two goals: to police inappropriate bias that might undermine sound scientific judgment and to support critical interchange of ideas among scientists. Therefore, these are the two greatest threats to methodological objectivity:

a. bias due to non-science influence on scientists (*individual bias*)

b. bias due to lack of criticism and debate among scientists (*community bias*).

Arguments supporting the conclusion that scientific advocacy will undermine scientific objectivity fall into two categories which mirror the two types of threat to methodological objectivity. One set of arguments stems from a concern that advocacy is likely to bias a researcher's individual judgment to the extent that she cannot maintain the impartial and independent perspective required to do research (*the Individual Bias Thesis*). The second set of arguments hold that scientific communities (whether research teams, university departments, professional organizations, or entire scientific fields) may lose their ability to perform objective research should they engage in advocacy (*the Community Bias Thesis*).

## Scientists make value judgments

The norm of objectivity in science has come under sustained philosophical scrutiny.[4] To take one particularly trenchant critique, Hugh Lacey has analyzed objectivity as composed of more particular requirements for impartiality, neutrality, and autonomy.[5] According to Lacey, *impartiality* requires that judgments in evaluating evidence and assessing the support for theories be based on cognitive, not social, values. That is, inferences that concern the strength of scientific evidence and how to interpret data ought to be supported by cognitive values like truth, reliability, and simplicity. Political and ethical values have no bearing on assessing the relevance or strength of evidence for a hypothesis. *Neutrality* requires, in essence, that the acceptance of a scientific theory is compatible with "all viable value complexes" and that acceptance of a scientific theory does not first require adopting a specific social, moral, or political perspective.[6] Specifically, the neutrality requirement rules out the possibility that some form of physics may warrant acceptance by one nationality but not another. *Autonomy* is the requirement that scientific practices, communities, and institutions must be isolated from any external interference which could mislead scientists from basing their inferences solely on data, cognitive values, and the internal direction of the scientific process.[7] Unlike impartiality and neutrality, Lacey finds little theoretical support and no practical support for autonomy. He concludes that, because scientists, scientific communities, and scientific institutions have always been embedded in, and worked cooperatively with, other social groups and institutions, this requirement is untenable.

Lacey thus holds that it is both possible and desirable for objective scientific reasoning to be impartial and neutral. Still, his account does not hold that values play absolutely no role in the judgments that scientists make. Because science is often relevant to important social issues—it is not, in other words, autonomous—it is legitimate for scientists to assess value commitments when they make decisions about which research questions to examine. Values also play a role when scientific knowledge is applied to develop new technologies, policies, and practices. For instance, Lacey argues that when deciding whether to invest limited resources in either agroecology or in biotechnology, it is desirable to consider which of these possible research programs is more compatible with economic and social goals. Doing so requires explicit discussion of values with regard to how to direct research and with regard to the value of the technological applications which research findings might support.[8]

To summarize, Lacey's position is that while:

> a. it is possible to conduct scientific research without making judgments which rely on social, political, or moral values,

and

> b. the inclusion of non-cognitive values in the process of scientific reasoning weakens the justification for any claims which follow,

it is nonetheless the case that

> c. advocacy work does not *necessarily* undermine the objectivity of science, so long as value judgments do not affect the interpretation and assessment of evidence.

Others agree with Lacey that values play an unavoidable role when science comes into contact with policy and practice, but they disagree with him about whether evidential reasoning can always be impartial and neutral.

For example, Heather Douglas' analysis of inductive risk demonstrates that in at least some scientific contexts, judgments about the risks of being wrong influence how to interpret data and how to assess the strength of evidence. This occurs when non-cognitive values are consulted in making decisions such as which data points to exclude from analysis, which statistical model to use to analyze data, how to classify observations, and how to structure null hypotheses and controls. When ethical values indicate that there is an unacceptable risk to, say, undercounting samples of a particular kind, then that value judgment will support a result which risks overcounts rather than undercounts.[9] In these sorts of contexts, where the relevance of research to ethical and social problems is obvious, it may be the case that decisions about research design are inevitably congruent with some value or other.

Helen Longino pursues a different strategy to show how values may unavoidably influence scientific reasoning. She argues that the underdetermination of theory by evidence has the implication that when a scientist makes a choice between two theories that are equally well supported by empirical evidence, the choice ultimately depends on implicit value judgments. Such value judgments typically rely on cognitive preferences, such as for simpler theories over more complex theories, but they sometimes incorporate non-cognitive values.[10] This is not necessarily an illegitimate use of values. Rather than substituting values in the place of empirical evidence, which *would* constitute an illegitimate form of bias, it may be a rational way to supplement inadequate evidence.

Thus, both Douglas and Longino have cast doubt on whether it is either possible or desirable, in all areas of scientific research, to avoid implicit ethical and social value judgments. They conclude that since it may be impossible and undesirable to avoid value judgments entirely, it is vital that scientists make their value-based reasoning explicit. This permits the scientific community to assess the role that values have played and to judge whether the use of values is justified.

Some scholars have extended the arguments of Douglas and Longino to show that if unexpressed and unsupported values have played a negative role in some scientific reasoning, then expressed and shared ethical values may likewise play a *positive* role. For example, feminist philosophers of science have developed case studies showing how values which explicitly support women's political interpretations of their situations (in anthropology) or which emerge from a feminist awareness of women's lives and biological differences (in archaeology and medicine) may assist in the development of scientific accounts which are more empirically adequate or better-reasoned than alternatives. [11]

In sum, there is an emerging consensus that it is impossible for scientific reasoning always to be free from social, moral, and political value judgments.[12] There continues to be controversy over whether values may play a positive role in scientific reasoning or whether it is best for scientists to reason only in ways that are neutral with regard to various value systems.

To be sure, this relatively recent consensus that science is not value-free does not settle the question of whether it is legitimate for individual scientists to act as advocates. There is still the concern that impassioned political commitments may obscure an individual's impartial judgment of the evidence.[13] What this consensus does show is that advocacy is not *necessarily* incompatible with scientific reasoning. I will later argue that it is not the act of publicly advocating for a policy that threatens methodological objectivity; it is the role that political preferences play in scientific judgment. And since scientists who are not advocates may have political commitments, too, public expressions of value commitments are not the litmus test that can distinguish biased from unbiased work. What matters to the validity of reasoning is not the existence of value commitments in the minds of scientists, but whether and how those values affect the execution and communication of research.

## Applied science is normative science

Some sciences are more likely to incorporate value-based assumptions into their research than others. The practitioners of these sciences share (or are expected to share) a normative viewpoint, or a set of shared values. In general, applied sciences share goals that extend beyond the cognitive goals of so-called "pure" or theoretical research. In addition to describing natural phenomena, the work of applied science is to bridge theory and practice. To take some examples, medical research is directed toward improving human health. Agricultural research is directed toward improving crops. Materials science, as well as other areas of engineering, are directed toward developing useful and safe technologies. Environmental sciences unify disparate scientific specializations in chemistry and biology under the shared normative goal of improving environmental health.

The core values which direct research in applied sciences often go without mention because they are widely shared, not only among members of the discipline, but also within society at large. Specific results in these sciences are considered valuable when they further a core value such as human or environmental health. The use of science to society drives research in applied disciplines. When publishing their findings, it is common practice for researchers to describe their work's relevance to a particular policy or technology. For this reason, it should not be surprising that in a normative discipline like environmental science, one study found that 94 percent of research articles make normative judgments and 55 percent state a policy preference.[14]

Conservation biologists have perhaps given more and deeper thought to the question of advocacy than practitioners in other applied sciences. They labeled their discipline as normative from its inception in the mid-1980s because conservation biology differs from related work in ecology, zoology, and botany due to its focus on biodiversity conservation.[15] As early as 1996 the journal *Conservation Biology* published an article by environmental scientist Dwight Barry and philosopher Max Oelschlaeger supporting conservation biologists' positive duty to advocate for environmental policies that protect species diversity. They argued that value considerations, when they are relevant to the goals of science, do not violate scientific norms for objectivity. Their position provides support for conservation biologists who wish to speak publicly in favor of protecting endangered species, restoring wetlands, curbing development and pollution, restricting the spread of invasive species, and related policy issues.[16] Although Barry and Oelschlaeger do not argue that each and every conservation biologist has the duty to engage in advocacy work, their position implies that the discipline as a whole does have this positive duty. This would mean that some substantial number of conservation biologists should engage in public work to communicate research findings, make the connections between research and policy, and become public advocates for responsible conservation.[17]

Reception of Barry and Oelschlaeger's article was mixed. In the same issue of *Conservation Biology*, Kristin Shrader-Frechette, a philosopher of science, extended the original argument in favor of advocacy by pointing out that if scientists do not take part in policy debates about matters on which they are expert, an interest in the common good based on scientific knowledge may well give way to less desirable bases for action.[18] Three conservation biologists also responded positively, agreeing with Barry and Oelschlaeger that advocacy work by conservation biologists need not undermine their scientific credibility. Reflecting the opinion of some of the founders of the discipline, these responses emphasized that the research findings of conservation biology, though scientifi-. cally rigorous, are intertwined with normative goals and policies. These scientists supported more explicit discussion of the content of values, of how values influence the science of conservation biology, and of how conservation biology can influence society.[19]

At the same time, three conservation biologists criticized Barry and Oelschlaeger's conclusions. C. Richard Tracy and Peter Brussard argued that there is a clear line between political work, based on values, and scientific work, based on facts. They raised the possibility that if policy were a subject for debate in the journal *Conservation Biology*, it would bring that professional publication into the same category as magazines published by organizations such as the Sierra Club. Scientific advocates could then find themselves in a position where "unsubstantiated dogma is substituted for scientific rigor."[20]

Finally, Earl McCoy argued that some of the activities which might be labeled advocacy are acceptable and even necessary to perform as science is currently practiced. However, he also raised the concern that discussions of values in conservation biology might lead scientists to allow value commitments to take the place of evidence-based scientific inquiry. McCoy made a distinction between the role that values play in the production, dissemination, and communication of scientific work versus their role as expressions of policy preferences that are outside the immediate work of scientific research. Only the latter counts as advocacy, for McCoy, and this role is politically involved and questionable. He asked

> Who could argue that explaining our research, using our scientific understanding to educate others, and even making practical recommendations based on our findings are inappropriate scientific undertakings? Scientists do these things regularly, when they write discussion sections in their research papers, submit final reports for contractual work, give talks to civic groups, and engage in many other activities. Explanation and informed speculation do not necessarily constitute advocacy.[21]

Still, McCoy does believe that scientists should not make controversial statements about policy positions that rely on their credentials as scientists. To do so, he claims, undermines not only their credibility as individual scientists but also the credibility of the profession.

## Not all forms of advocacy are controversial

Until now, I have written as though advocacy activities are all of a piece. The debate about whether the objectivity of science is at odds with advocacy glosses over the distinction between types of advocacy, and participants sometimes assume that all forms of advocacy will stand or fall together. However, as we just saw, McCoy raises the question of exactly which activities fall under the umbrella of advocacy. Likewise, the philosophical scholarship finds that while values do play a legitimate role in the direction and application of science and may further play a positive role within scientific reasoning, they may also be misapplied in a way which undermines the objectivity of research. Rejecting the model of science as value-free blocks the construction of a simple and firm division between the descriptive activities of scientists and the prescriptive activities of policy-makers. As we have seen, values are already incorporated into science, and scientists are essential to forming responsible policy. It is clear that not every use of normative language by a scientist is inappropriately subjective, yet it is equally clear that bias is a real risk for scientists who are in positions to influence policy. The task at hand is to clarify when normative judgments are innocuous, unavoidable, or justified, and when advocacy is best characterized as inappropriate bias.

Making this distinction clear will require that we distinguish various forms of advocacy. Analyzing the legitimacy of particular cases of advocacy will depend on examining the content of non-cognitive values, the extent to which they are shared within and without scientific communities, and whether the content of values is open to critique and revision. Following a case-by-case analysis of examples of scientific advocacy, I will answer objections from those who continue to worry that advocacy will undermine scientific objectivity or scientific credibility, and I will suggest a series of issues that remain to be addressed.

The first point to make is that, obviously, scientists may act as advocates while not in their professional role. That is, scientists may join environmental organizations just as they might join the ACLU or volunteer to make calls for a political party. Whether scientists' advocacy activities should be limited in ways that others' activities are not is not at issue. Likewise, professional organizations may engage in lobbying for increased federal funding of science or may support campus labor unions, but these political activities have no bearing on the production of their scientific research. What *is* at issue is whether it is appropriate for scientists to use their credentials *as* scientists when engaging in advocacy.

There are a variety of types of advocacy activities that scientists engage in. Within these roles, scientists rely on their professional credentials, and we should evaluate whether these roles may either bias their research or undermine their credibility. In each case below, a scientist or group of scientists either use normative or value-laden language, or stipulate a policy or management preference:

A. an aquatic biologist learns that her university is planning a development project on campus which will violate the spirit and perhaps also the letter of federal wetland regulations, and she writes a letter to the university president expressing her concern;

B. in college classes, an environmental science professor teaches that climate change poses an urgent threat to human well-being and to biodiversity;

C. an environmental toxicologist accepts speaking engagements to a variety of audiences on the topic of environmental cancers;

D. the same environmental toxicologist also testifies in legal cases;

E. a marine ecologist is asked by a local newspaper reporter for his opinion on the planned eradication of an invasive species of marsh grass, and he gives it;

F. a scientist serves on a project review board for a funding agency and, following the instructions of the funding agency, her recommendations are based both on the quality of the research proposal and on its social relevance;

G. a scientific society asserts that biological diversity is inherently good;

H. a university department is offered a lucrative contract by a private firm, providing support for research and education but requiring, in turn, that faculty members obtain consent from the company in each instance that they wish to communicate publicly using normative language or in relation to policy preferences.

Let us examine these in turn. I will suggest that there is an important difference between cases A – C and cases D – G. Case H is obviously an outlier relative to the other cases, and I will discuss why this form of control of scientists' advocacy activities constitutes a barrier to their performance of professional obligations.

The biologist in Case A is clearly engaging in advocacy by contacting her university's president and expressing a political view that's both outside of her required scientific work and independent of her responsibilities as a faculty member. She writes a letter to a policy-maker with the intention of influencing a decision. Her scientific expertise puts her in a position to be aware of the ecological value of the wetland and of the federal regulations; however, it is unlikely that writing such a letter would bias her research. The concern that engaging in this form of advocacy would bias her reasoning ability is misplaced. Moreover, the value for wetland conservation, which she supports, is widely shared among ecologists and in the community at large and is further supported by federal law. (Indeed, her concern may be motivated by loyalty to the university and a desire that her institution avoid wrong-doing and hefty federal fines.) This action does not undermine the objectivity of her scientific work or her credibility as a scientist; it may even be required by her sense of professional integrity.

In Case B, the scientist's actions in the classroom are an expected part of the job but not integral to research. Some might argue these actions go beyond a strict definition of teaching the facts of environmental science. However, since environmental science embraces teaching the consequences for human health and well-being of environmental change, and since the physical evidence supporting climate change and the social science evidence supporting its negative effects is well-established, this expression of values also seems to be well within the bounds of what is legitimate for environmental science faculty. Indeed, an environmental scientist would *lose* credibility for teaching topics such as climate change without mentioning the negative impact on biodiversity preservation and human well-being.

The toxicologist's actions in Case C resemble those of the professor in Case B. Both involve a researcher communicating the relevance of an area of science to human health and well-being. The professor's actions in case B were justified because communicating the findings of climate science and their policy implications is a central part of teaching environmental science. Likewise, it matters in Case B that the values expressed are widely shared. Similarly, in Case C we would assume that the toxicologist values human health and fighting cancer. However, we could certainly conjure up a case in which the toxicologist is skeptical about environmental causes for cancer and has been convinced of this by his research work, even though there is general agreement in the research community that many cancers have an avoidable environmental trigger (e.g., by wearing sunscreen one is at less risk for skin cancer). In contrast, there is room to be concerned about the objectivity of an environmental toxicologist who dismisses, for example, the dangers of dioxin to human health. The problem this case raises is not whether it is acceptable for scientists to communicate to the public per se, but whether the scientific knowledge they convey is well-established and backed by the scientific community and whether the values that back it are widely shared in society.[22]

Case D involves the same toxicologist as case C but introduces the further complication that the concept of objectivity in science is not well-matched with the concept of objectivity in law. While objectivity in science indicates that researchers have impartially evaluated evidence and that their findings have been submitted to criticism by peers, objectivity in law is achieved through adversarial engagement by representatives of parties with conflicting beliefs. Scientists may testify about the general state of scientific knowledge, namely, what is accepted by the scientific community. Or they may be asked to testify concerning their own research or observations. In either case, concerns about inappropriate bias are amplified in the legal setting.[23] The court system is one route by which science is communicated and its findings integrated into policy. It would be a detriment to the public good to discourage scientists from communicating about science in this way. This form of advocacy is beneficial to the scientific enterprise and to the public, and yet it is especially risky because the stakes, and the potential for influence, are higher.[24]

Like Cases B, C, and D, Case E (the newspaper interview) describes a means by which scientists communicate to the public. This case, however, raises the most typical question of the appropriateness of advocacy. How general ought the ecologist's comments be? Is he limited to describing the effects that the invasive marshgrass has had on the estuary's ecology? Should he couch his comments in an explicit and broadly accepted value framework, perhaps saying that elimination of the invasive species is the best way to guarantee the survival of native species? Or may he also give an opinion on more controversial aspects of the project, such as which areas should be prioritized, the preferred method for control of the native species, and the ecological value of the eradication relative to its costs? What if eradication of the invasive species is required to stabilize the native ecology but will, at the same time, threaten rare or valuable species that have shown a preference for the introduced marshgrass?

This last question focuses attention not on whether advocacy is appropriate or not, but on whether advocacy is appropriate in controversial cases, which are exactly the cases where the scientific community is itself likely to be split and a scientist might be most tempted to advocate for a particular policy.

Among the tasks that scientists are expected to perform is peer review. In addition to reviewing articles for journals, many scientists work with funding agencies to review grant proposals. Since agencies such as the National Science Foundation routinely ask that applicants justify the value of their proposed research, assessing the relevance of scientific research to social goals has become a routine part of scientific practice. Some grants are targeted specifically at research that would elaborate on or further policy goals. This raises the question of whether a peer reviewer (Case F) should take her own policy preferences into account when judging such proposals. Or should she prioritize proposals which are value-neutral, assuming that value-neutral research can be identified? Or should she attempt to identify the most widely shared community values and accept proposals most relevant to those values? Whatever the answers to these questions, reviewing grant proposals is one way that scientists can actively work in support of "well-ordered science," or science that best serves the public good.[25]

Case G involves a scientific organization expressing its consensus on a normative statement which is relevant to public policy but which can be neither confirmed nor disconfirmed by scientific evidence. Unlike the normative goal of medicine to support human health, the value of biodiversity is philosophically vague and not universally accepted by the public.[26] Nonetheless, because the relevance of research proposals is judged relative to values, it is sensible and legitimate for the organization to make this kind of public statement to their members and to non-scientists.

Case H clearly describes a potential conflict of interest, and it alone among these examples leaves no question that the contract, if accepted, would negatively affect scientists' abilities to perform research which is relevant to improving public policy. The sharpest concern is not that the acceptance of private funding would bias researchers (though it may do that, too, cf. Douglas 2000). The sharpest concern is that it would prevent the publication of sound research. If we think of the core of scientific work as the production of descriptive knowledge, irrespective of value judgments, then any sort of ban or control on advocacy activities should make no difference to how scientists do their job, whether it is self-imposed or a condition set by funding agencies or donors. It is clear that this is not the case, however. It is easy to imagine how control over scientists' public communication about the implications of their research could develop into control over which research projects they take up. At the very least, the direction of these scientists' research programs would likely be shifted to research areas that would not threaten the interests of the firm they are contracted to. This case would be even more egregious, of course, if a contract with academic scientists dictated policy positions they were required to support.

# The content of advocacy matters

All of these cases, A – H, may be grouped under the rubric of advocacy. Not all of them are uncommon or elicit criticism within scientific circles. Some of these activities necessarily involve value judgments (at least in the applied, normative sciences such as environmental science), but they are at the same time common and even required tasks for academic researchers.

Cases A – C share the feature that they involve an individual acting in such a way that the expressed normative views are easily identifiable as the individuals' own. A scientist writing to a policy-maker to express an opinion, even when based on her research expertise, represents only herself (Case A). College professors are generally expected to leave their politics out of the classroom, but when this does not happen, college students can be expected to be mature enough to distinguish between the lesson and the commentary (though in Case B the political content is constitutive of the field's commitment to studying and improving environmental health). And a responsible speaker can clarify to an audience which views are personal and which are shared with the scientific community (Case C).

However, in cases D – G the scientists speak on policy issues in a way that represents the view of a scientific community. Scientists would not be asked to testify in a court or comment for a newspaper about mere personal opinion (Cases D and E), but only with regard to their scientific judgment, and scientific judgment is jointly produced within a community that evaluates the work of individuals. Individuals are asked to comment or to testify based on their reputations (reputations granted by their peers). Likewise, funding agencies (Case F) and professional societies (Case G) represent diverse communities of scientists and have a responsibility to represent the general consensus of their constituents while not repressing dissent.

The final case (H) illustrates a possible conflict of interest. The danger it highlights is the transfer of responsibility for assessing the scientists' advocacy from the scientific community to the hands of a corporation that does not share the scientific commitment to discovering natural truths and disseminating them in the service of the public good. The scientific community is in the right position to evaluate the content of its members' advocacy claims; other bodies are not.

McCoy's comments (in the section "Applied Science is Normative Science" above) illustrate the ambivalence many scientists have for advocacy work. On the one hand, some advocacy activities are routine, but on the other hand, visible and zealous political engagement may threaten the credibility of the profession. I suggest, however, that his response to this risk, that in a perfect world scientists would never undertake any explicitly political work beyond doing descriptive research, is an over-reaction. McCoy suggests that the response to potentially embarrassing zealots is to frown on advocacy, except where it is among the things that scientists do regularly.[27] But my examination of sample cases of advocacy reveals that the problem is not one of identifying which actions are permissible and which are not. The problem with zealots is that they advocate policy preferences that are not widely shared within their communities. Cases D (the toxicologist who may testify contra the general scientific consensus) and E (the ecologist who may publicly support a management policy which his colleagues are divided on) best illustrate this point. It is the *content* of advocacy statements that can make some of them problematic, *not* the type of action, its form or audience.[28]

# The risk of bias does not support a ban on advocacy

Like McCoy, Robert Lackey, a fisheries biologist, has argued that although it is appropriate and necessary for scientists to contribute to the formation of public policy by performing relevant research studies, by publishing their findings, and by doing what they can to encourage policy makers to pay attention to those findings, scientists must stop short of recommending specific policies. He describes the limits on what scientists may legitimately do as advocates:

> One common concern about the science-policy interface is that some so-called science is imbued with policy preferences. Such science is labeled as normative and its use is potentially an insidious kind of scientific corruption. What separates normative science from "regular" science is that normative science is developed, presented, or interpreted based on a tacit, usually unstated, preference for a particular policy . . . ..Attempting to be both the provider of policy-neutral science and an advocate for one's personal policy preferences is laden with conflicts of interest and . . .is potentially unethical.[29]

By being concerned about conflicts of interest and about research being policy-neutral, Lackey is referring to the common perception that good science should be value-free. He believes that science which is not value-free, even from the start of a research project, may end up producing biased results. Worse, he argues, is the fact that "Normative science often is not perceptibly normative to policy makers or even to many scientists."[30]

Lackey is concerned that by having a policy preference, a scientist is more likely to find evidence that supports that policy than evidence that calls it into question. In addition, values work as stealth bias: it is not necessary for the scientist to be *intentionally* slanting her results. She may be *unaware* of the bias that nonetheless corrupts her results. Lackey's criticism is the most pointed of those I have considered. He recognizes the role that values play in scientific reasoning, and he focuses on the development of conflicts of interest.

I have argued that applied sciences are influenced by values. Though this influence may be positive, it carries risks. Lackey's concern makes reference to the *individual bias thesis*: scientists will not be able to identify the point at which their own research becomes unacceptably biased. Further, he worries that scientists will not be able to recognize bias in their peers' work if they share the bias themselves. This, then, refers to the *community bias thesis*. Lackey and others have responded to individual and community bias with wariness about any form of advocacy. This response, however, is both untenable, given the constitutive role of values in applied science, and undesirable, given the degree to which scientific institutions currently rely on their members to carry out advocacy work.

## The control of bias lies with scientific communities

There is another alternative. The general response to the danger of individual bias has been to use the oversight of communities: research results, for instance, are reviewed by peers for the quality of their supporting evidence. The general response to concerns about community bias has been to foster criticism within the community and to respond to criticism from outside the scientific community. More specifically, the alternative that responds to both individual and community bias has been to formulate guidelines on professional behavior and conflicts of interest and to revise these guidelines, as needed, in response to criticism.[31] The debate about whether advocacy is at all appropriate in, for example, the environmental sciences, is ripe for a turn to more concrete discussion of particular norms for political involvement and conflicts of interest. Formulating guidelines to guard against conflicts of interest (as illustrated in Case H) should occur alongside a discussion of which forms of advocacy are most effective for communicating the relevance of research to policy and for creating the conditions for well-ordered science.

Other disciplines, particularly medicine and biotechnology, have formulated conflict of interest policies which should inform what is required of other normative, applied sciences.[32] The requirements of conflict of interest policies in different disciplines, however, will necessarily be different depending on the field and the forms of influence that are most likely to affect individual and collective judgment.[33] To take a concrete example, environmental science fields will likely be challenged by responses to the 2010 oil spill in the Gulf of Mexico. Reportedly, some marine scientists who work in the area of the spill have been offered lucrative consulting contracts by BP, and the terms of the contracts may prevent scientists from publishing their research, from discussing their research with the press or other scientists, and from testifying against BP in court. They may also be required to testify on behalf of BP.[34] A contract which limits the direction of academic research, limits what research can be published, and limits how it is communicated to the public represents control over scientific research and advocacy activities and acts as a form of advocacy itself.

Acceptance of an advocacy role for scientists raises further questions, such as how to effectively address public perceptions of bias, how to communicate across normative differences between scientists and the public, and what to do when scientific advocates conflict. These are non-trivial questions whose answers will no doubt evolve over time. For instance, do conservation biologists agree among themselves that biodiversity is intrinsically valuable, and should they work to support the development of policies based on this value in an international political context that is oriented around anthropocentric valuations? And what is proper professional behavior for scientist-advocates whose values are in conflict? For instance, clashes between biologists who advocate for biodiversity conservation in Africa and social scientists who advocate for linguistic and cultural preservation have affected the politics of conservation as well as their own research.[35] How can interdisciplinary work be moved forward when scientific research programs are oriented around different normative goals?

Contemporary philosophical accounts of objectivity in science accept a role for value judgments. Therefore, the very fact that advocates have value commitments does not cast doubt on the rigor or objectivity of their work and does not distinguish them from scientists who hold value-based commitments but do not engage in advocacy work. There may be practical reasons for some scientists, sometimes, to avoid engaging in advocacy, and there are also some types of advocacy that should at least remain suspect. But most forms of advocacy do not pose a principled threat to scientific objectivity.

# Notes

1. David Hull's *Science as a Process* (Chicago: University of Chicago Press, 1988) demonstrates why studying the practices of scientific communities (such as peer review, priority disputes, and the formation of professional societies) is subject matter for philosophy of science. Some institutional structures and community practices facilitate the production and dissemination of scientific knowledge while others hinder it. Thus, the progress of science has been not just a matter of improving scientific instruments or methods but also of constructing social practices which facilitate critical knowledge production.

2. The journal *Conservation Biology* recently published a special section on "Policy Advocacy and Conservation Science," and the editor notes in his introduction that after

more than a decade of intense debate in the journal and other professional venues, "the issue of advocacy by scientists in public policy has been addressed frequently, but with no consensus resolution" (Meffe, Gary K. 2007. Policy Advocacy and Conservation Science: Introduction. *Conservation Biology* 21(1): 11).

3. For more complete analyses of the multiple modes of the objectivity concept in philosophy of science, see Lloyd, Elisabeth A., "Objectivity and the Double Standard for Feminist Epistemologies." *Synthese* 104, no. 3 (1995): 351-381 and Douglas, "The Irreducible Complexity of Objectivity." *Synthese* 138, no. 3, 2004: 453-473.

4. Analyses of scientific objectivity do not all share the same definitions of terms nor take up problems from the same angles, and the literature is now extensive. I focus on a few accounts, but many more views can be found in anthologies such as Carrier (2008), Kincaid, et al. eds., *Value-Free Science?: Ideals and Illusions* (New York: Oxford University Press, 2007) and Machamer et al. eds., 2004. *Science, Values, and Objectivity* (Pittsburgh: University of Pittsburgh Press, 2004).

5. Lacey, Hugh. *Is Science Value-free?: Values and Scientific Understanding* (New York: Routledge, 1999).

6. For the sake of space, I present a much simplified version of Lacey's analysis of neutrality. For the full account, see Lacey, *Is Science Value-free*, 80.

7. Lacey. *Is Science Value-free*, 255.

8. Lacey, Hugh. Assessing the Value of Transgenic Crops. *Science and Engineering Ethics* 8, no. 4 (2002): 497-511.

9. Douglas, H. "Inductive risk and values in science." *Philosophy of Science* 67, no. 4 (2000): 559-579.

10. Longino, Helen. *The Fate of Knowledge* (Princeton: Princeton University Press, 2002), and "How Values Can Be Good for Science." Machamer et al. eds., 2004. *Science, Values, and Objectivity* 127-142.

11. See Clough, this volume; Crasnow, Sharon. "Activist research and the objectivity of science." *American Philosophical Newsletter on Feminism and Philosophy* 6 no. 1 (2006): 3-5; Wylie, Alison and Lynn Hankinson Nelson. "Coming to Terms with the Values of Science: Insights from Feminist Science Scholarship." Kincaid, et. al. eds., in *Value-Free Science*, 58-86.

12. So far as I know, no philosopher claims that all science incorporates non-cognitive value judgments all the time. However, it appears to be more likely that values will come into play when science has implications for policy, and it is these cases which are at stake when we consider the legitimacy of scientific advocacy.

13. If anything, this concern is strengthened rather than placated by the argument that values may play a positive role in scientific reasoning. This is because that argument rests on a comparison between examples of scientific reasoning which incorporate liberatory values and those which entail exclusionary political values. The argument supporting a positive role for values shows that it is the content of value beliefs, not the fact of their existence, which matters.

14. Scott, J. Michael, et al. "Policy Advocacy in Science: Prevalence, Perspectives, and Implications for Conservation Biologists." *Conservation Biology* 21 no. 1 (2007): 29-35) examined six flagship journals in natural resources disciplines for: "use of normative or value-laden language and stipulation of a preferred policy or management preference" (30).

15. Michael Soulé's description of the field's founding identified its normative postulates and the policy role that conservation biology would play: "Biologists can help increase the efficacy of wildland management; biologists can help improve the survival odds of species in jeopardy; biologists can help mitigate technological impacts" (Soulé 1985, 733).

16. Barry, Dwight and Max Oelschlaeger. "A Science for Survival: Values and Conservation Biology." *Conservation Biology* 10 no. 3 (1996): 905-911.

17. Barry and Oelschlaeger address conservation biologists as a group when they write "[W]e have an ethical obligation to provide decision makers with explanatory knowledge and prescriptive recommendations . . .Without providing prescription along with knowledge gained by our studies, we will have no basis for complaints that our knowledge is ignored in the policy arena" (1996, 910).

18. Shrader-Frechette, Kristin. "Throwing out the Bathwater of Positivism, Keeping the Baby of Objectivity: Relativism and Advocacy in Conservation Biology." *Conservation Biology* 10 no. 3 (1996): 912-914.

19. Maguire, Lynn A. "Making the Role of Values in Conservation Explicit: Values and Conservation Biology." *Conservation Biology* 10 no. 3 (1996): 914-916; Meine, Curt, and Gary K. Meffe. "Conservation Values, Conservation Science: A Healthy Tension." *Conservation Biology* 10, no. 3 (1996): 916-917

20. See Tracy, C. Richard and Peter F. Brussard. "The Importance of Science in Conservation Biology." *Conservation Biology* 10, no. 3 (1996): 918. The philosophical justification for this position has to be the belief that objectivity can only be based on value-freedom and that the alternative is pernicious relativism. However, the philosophical work cited in Section 1 above gives a subtle and thorough account of how to produce scientific objectivity out of work which interconnects with value commitments. This social turn in philosophy of science also explains why it is the case that so much of science has been epistemically reliable even when researchers' highest consideration is the advancement of their own career, not the advancement of science or the public good (Hull 2001).

21. McCoy, Earl. "Advocacy as Part of Conservation Biology." *Conservation Biology* 10, no. 3 (1996): 920.

22. This concern might vanish if we were told that the scientist represented himself to his audiences as a renegade and his ideas as dissent from the mainstream.

23. Foster and Huber, *Judging Science: Scientific Knowledge and the Federal Courts* (Cambridge: MIT Press, 1999) examine the rules that cover testimony on scientific matters of fact and analyze the ways that the legal rules that govern scientific testimony conflict with the understanding of scientific evidence and communities that emerges out of philosophy of science.

24. Sociologist Steve Fuller's testimony in *Kitzmiller et al. v. Dover Area School District* is a case in point. Fuller has been criticized for not taking seriously the political import of his legal testimony in support of intelligent design creationism on the basis of his postmodern theories about science (Lynch 2006).

25. Kitcher, Philip. Responsible Biology. *BioScience* 54, no. 4 (2004): 331-336.

26. Sarkar, Sahotra. *Biodiversity and Environmental Philosophy.* (New York: Cambridge University Press, 2005).

27. McCoy, Earl. "Advocacy as Part of Conservation Biology." *Conservation Biology* 10 no. 3 (1996): 919-920.

28. Nelson and Vucetich, "On Advocacy by Environmental Scientists: What, Whether, Why, and How." *Conservation Biology* 23, no. 5 (2009): 1090-1101 survey and respond to a broader array of scientists' arguments against advocacy than I examine here. They, too, find the arguments against advocacy to be inconsistent or unconvincing.

29. Scott, J. Michael, et al. "Policy Advocacy in Science: Prevalence, Perspectives, and Implications for Conservation Biologists." *Conservation Biology* 21 no. 1 (2007): 29-35.

30. Scott, J. Michael, et al. "Policy Advocacy in Science," 29-35.

31. Intemann and de Melo-Martin (this volume) discuss the importance of revising conflict of interest policies as the effectiveness of policies becomes better understood and in response to changes in the legal and institutional environments in which scientists work.

32. For references on conflict of interest policies in the biomedical sciences, see Intemann and de Melo-Martin (this volume).

33. One tends to think of questionable behavior in biotech as resulting from the lure of high profits, while a high-profile environmental scientist who has gained notoriety for his political advocacy is James Hansen. Many environmental scientists admire Hansen's involvement in political protests against coal-mining because they see it as in the interest of the public good. Nevertheless, it is time for an explicit self-appraisal within various communities with regard to the limits of political advocacy.

34. Nelson, Cary. "BP and Academic Freedom." *Inside Higher Ed.* 22 July 2010. http://www.insidehighered.com/views/2010/07/22/nelson (accessed August 20, 2010).

35. Agrawal, Arun and Kent Redford. "Conservation and Displacement: An Overview." *Conservation and Society* 7, no. 1 (2009): 1-10.

# Chapter 12

# Bias, Impartiality, and Conflicts of Interest in Biomedical Sciences

## Kristen Intemann and Inmaculada de Melo-Martín

When Congress passed the Bayh-Dole Act in 1980, ties between industry and academic researchers were relatively uncommon.[1] Bayh-Dole encouraged universities and individual researchers to patent and commercialize inventions that resulted from federally funded research. The new law led academic institutions to set-up technology transfer offices, triggered the creation of start-up companies, and promoted private funding for clinical studies. As a result, the number of U.S. patents obtained by universities increased almost sixteen-fold between 1980 and 2004.[2] There are about 300 technology transfer offices, nearly a ten-fold increase from before the enactment of the law, and more than 4,500 for-profit firms.[3] With 58 percent of the funding, pharmaceutical, biotechnology, and medical device companies supply the largest proportion of total research spending.[4]

There is disagreement about whether Bayh–Dole has truly fostered new discoveries, contributed to bringing more biomedical products to the market, or increased university profits.[5] It is clear, however, that the commercial activities encouraged by the Bayh-Dole Act create conflicts of interests for both individual researchers and institutions and thus have the capacity to bring bias into the scientific research enterprise. For instance, a variety of studies have found that research receiving funding from commercial parties is significantly more likely to report positive outcomes than studies funded by not-for-profit organizations.[6] While the term "conflict of interest" is used in a variety of ways, there some agreement that a conflict of interest is a situation in which a financial or some other personal consideration (the secondary interest) has the potential to bias, or appear to bias, professional judgment and hinder objectivity (the primary interest).[7]

In 1995, the Public Health Service, mindful of the fact that the proliferation of financial interests and the increased presence of industry-academic partnerships could bias, or be perceived to bias, researchers' judgments when conducting or publishing their investigations, issued its first conflicts-of-interest regulation (42CFR50-Subpart F). It required grantees to report to their institutions significant financial conflicts (more than $10,000 per year from a given company, or 5 percent equity in a company) that are related to the research. Similarly, institutions have the responsibility to ensure that researchers comply with disclosure requirements and have specific policies to manage, reduce, or eliminate such conflicts of interests. The stated goal of these regulations is to promote "objectivity in research by establishing standards to ensure there is no reasonable expectation that the design, conduct, or reporting of research funded under PHS grants or cooperative agreements will be biased by any conflicting financial interest of an investigator." (42CFR50-Subpart F)

Because of increased concern with the effect of financial interest on the integrity of scientific research, the AAMC (2001), the AAU (2001), and more recently the IOM (2009) have called institutions to strengthen their conflicts of interest policies by requiring investigators to disclose all payments that are directly or indirectly related to their research.[8] These organizations also agree that overall central goal of conflict of interest policies is to protect the objectivity of scientific investigations. And they also agree that such objectivity is threatened by conflicts of interest because such conflicts conceivably can impart bias in the conduct or report of their research.

Although there has been a significant amount of discussion on the need of conflict of interest policies, reflections on whether they can achieve their stated goals are scarce.[9] In this chapter we attempt to correct that problem. After rejecting some of the arguments that have been offered criticizing conflict of interest policies, we present the different types of polices that have been advanced to deal with conflict of interest in biomedical research[10] and assess their effectiveness in achieving their espoused aims. We conclude that conflict of interest policies are ineffective in attaining these goals because they rely on overly simplistic conceptions of bias and impartiality. As a result, they do very little to prevent bias or promote impartiality in biomedical research. Seeing how such policies fail, however, provides insight into how bias operates as well as how objectivity might be better promoted.

# Criticisms of Conflict of Interest Policies

Conflict of interest policies have been denounced for a variety of reasons. Most of these arguments, however, do not consider whether the policies in fact promote their stated aims. As a result, such arguments tend to criticize conflict of interest for failing to do something they are not intended to address. For example, some have criticized financial conflict of interest policies for narrowly focusing on financial conflicts and neglecting other types of conflicts—desire for fame, need to get grants—that can also bias researchers' professional judgments.[11] Yet the fact that other conflicts may be a problem does not mean that policies regulating financial conflicts are not needed or are ineffective. Similarly, some have criticized conflict of interest policies for failing to prevent intentional bias, or cases where scientists manipulate data, skew statistical analyses, or misrepresent research findings in order to serve their financial interests.[12] While it is true that disclosure policies and oversight committees are unlikely to identify or prevent such behavior, it is not clear they are intended to do so. In fact, this is why there are regulations against research misconduct in addition to conflict of interest policies. Finally, some have argued that conflict of interest policies fail to address how the commercialization of research can affect which research projects or hypotheses are developed and investigated.[13] Pharmaceutical research money, for example, is often used to study health problems that affect wealthier populations and are more lucrative (such as erectile dysfunction or depression) rather than health problems that are particularly serious or prevalent in developing countries (such as AIDS or malaria). While this is clearly a problematic consequence of the commercialization of research, it is not obvious that conflict of interest policies are intended to address this particular problem.

Another common criticism of conflict of interest policies is that they have potentially bad consequences for science. For example, some have argued that such policies call into question the integrity of honest researchers or that they prevent certain kinds of research from being pursued and thus constitute a form of censorship.[14] Even if there were empirical evidence for these claims, this would only establish that the policies have some undesirable consequences. But many policies are adopted because the benefits outweigh the costs. In other words, additional arguments would be necessary to show that conflict of interest policies do not have benefits that are sufficient to outweigh their costs.

Finally, some policies, such as requiring scientists to disclose their financial interests to universities, funding agencies, institutional review boards, journals, or professional associations, have been criticized as ineffective because scientists often do not comply.[15] This, however, seems to be a concern about how to enforce conflict of interest policies, rather than a problem with the policies themselves.

Thus, while conflict of interest policies have been condemned for a variety of reasons, better arguments are needed. In particular, it is necessary to evaluate whether such policies are effective in promoting their stated aims, at least when scientists comply.

## Conflict of Interests Policies and their Goals

Research institutions, federal agencies, and publishers have all developed policies that attempt to control the negative effects of financial conflicts of interest. We have dealt with the inadequacy of conflicts of interest policies adopted by journals elsewhere.[16] So, in this chapter we focus on regulations developed by research institutions.

Also, our main concern is on policies directed to individuals rather than those that address institutional conflict of interest. Individual scientists may be subject to policies drafted by different institutions. For example, one might be regulated by university policies in addition to the policies of government agencies when conducting federally funded research, as well as policies of journals when publishing research results. Though these different institutions all have conflict of interest policies, the particulars of such policies can vary significantly.[17]

Institutional policies governing the financial conflicts of individual scientists also vary greatly across the nation.[18] Following federal guidelines and recommendations by organizations such as the AAMC, AAU, and the IOM, research institutions have tended to develop conflict of interest policies that include the three distinct, but related, elements: disclosure of conflicts; management of those that are thought to be significant; and prohibition of research activities when such is thought to be necessary to protect the public interest or the interest of the university.[19] In the following sections we briefly describe these different elements, the goals that they are designed to accomplish, and how they fail to promote those goals.

# Disclosure

The requirement to provide information about funding sources, honoraria, consultancies, stock ownership, patents applications, and royalties is a common element of conflict of interest policies. A significant number of medical centers have the $10,000 as threshold over which investigators are required to disclose certain financial interests to their institution.[20] Other schools have a lower threshold or might require disclosure of all sources of income no matter how small. Also, institutions might require disclosure not only to the institution but also to other stakeholders, e.g., research subjects, or journals.[21]

Generally, institutions have attempted to develop disclosure policies that are clear, easy to comply with, and that obtain sufficient information about the nature, duration, and scope of the financial relationship, so as to permit institutions to determine the risk of biasing researchers' judgments when conducting or publishing their investigations.

One stated justification for disclosure requirements is that they can prevent bias by discouraging researchers from entering into financial agreements that might inappropriately affect their professional judgments.[22]

However, it is not clear that researchers find financial agreements to be a problem, at least not to their own integrity.[23] Some studies suggest that they find conflict of interest policies irrelevant, an imposition on individual self-determination, or counter-productive.[24] Criticisms against NIH policies that prohibit their researchers from entering into a variety of such agreements confirm the commonality of such beliefs.[25] Similarly, many investigators believe that by enforcing those policies the NIH will have difficulties recruiting top scientists, which also suggests that they do not find the possibility of conflicts of interest to be a serious concern.[26]

Moreover, as disclosure of financial conflicts becomes common and widespread, such conflicts risk being seen as "normal," rather than anything that should be the source of shame. As mentioned earlier, there has been a proliferation of academic and industry partnerships, not a curtailment. Thus, there is currently no evidence that disclosure policies do much to discourage scientists from entering into arrangements that might lead to financial conflicts of interest. Rather, such conflicts appear to increasingly be accepted as commonplace.

Disclosure requirements might conceivably also help prevent bias by making researchers more attentive to the ways in which financial interests might influence their behavior when conducting research. Knowing that they have to declare conflicts of interest and that their research will be under significant scrutiny, researchers could be more attentive to following good scientific practices and methodologies.

Yet there is little empirical evidence that this is the case and some evidence that most researchers are so confident that their scientific judgments are not influenced by financial interests that they do not need to be any more attentive than they otherwise would be.[27] More importantly, this justification assumes that researchers will be able to "catch" themselves if financial interests do influence their research. But, presumably, a significant danger of financial interests is their ability to lead to *unconscious* bias. That is, even well-intentioned scientists may inadvertently allow their financial interests to influence how research questions are framed, which methodologies are preferred, and how data is selected and interpreted. Such biases are nearly impossible for individual scientists to identify in their own research. This is so precisely because they occur unintentionally, even when conscientious scientists believe they are rigorously adhering to the accepted methodologies of their field.[28]

Moreover, some studies also suggest that disclosure may in fact exacerbate the problems that result from conflicts of interest. This is so because those with competing interests might be less concerned about bias once they know that relevant stakeholders had been warned of those conflicts.[29] Thus, they might feel they do not need to be as careful as they might otherwise be.

A third justification for disclosure policies is that they can help others identify biases when they do occur. Such requirements might be thought to alert IRBs, conflict of interest committees, journal editors, reviewers, readers, and potential clinical trial subjects to the possibility that the interests of the researcher may have biased the framing of the research question, the choice of methodologies, assessment of risks, statistical calculations, or the ways risks of participating in a clinical trial are presented. As a result of this increased attention, greater scrutiny can be applied to the researcher's study design, methodological choices, and the assessment or presentation of risks.

This justification assumes, however, that disclosure of financial interests helps enable others to better identify when those interests have inappropriately influenced research. When bias occurs, however, it either manifests in an obvious breach of scientific norms or in more subtle ways. If financial interests result in a fairly apparent breach of scientific norms, this is likely to be identified by IRBs, editors, or reviewers regardless of whether financial interests are disclosed or not. In this case disclosure policies do not do any work in identifying the bias.

More likely, however, financial interests bias research in more subtle ways, such as unconsciously affecting implicit background judgments regarding the characterization of data, selection of evidence, or the sorts of risks that are ac-

ceptable to human subjects. Yet these types of judgments are not always transparent to IRBs, clinical trial subjects, editors, reviewers, or the public. Consider,
for example, studies funded by chemical companies that purport to show that
low-levels of environmental toxins can actually have some beneficial health effects. One such study presents evidence that low doses of cadmium significantly
shrink testicular tumors in rats (Cook and Calabrese 2006). The fact that a toxin
may have beneficial effects with respect to *some* biological endpoint, however,
does not show that it is beneficial with respect to *all* (or even the most important) biological endpoints (Elliott 2008b). For instance, a dose of cadmium that
is effective in shrinking testicular tumors may also be likely to produce brain
tumors or heavy metal poisoning in the same subjects. In this sort of case, financial interests could influence the selection of endpoints and the framing of research data. But again, the role that financial interests have in these sorts of decisions will not always be apparent to anyone evaluating the research. Those
assessing the research may know that there are other biological endpoints that
would be relevant. Clearly, simply scrutinizing research to ensure that scientists
adhere to scientific norms, methods, and practices is unlikely to determine
whether the research is biased or not.

If the judgments that have been influenced by financial interests are not apparent, then it will be difficult for others to identify bias, regardless of whether
financial interests are disclosed.[30] Indeed, the fact that commercially funded
studies are much more likely to report positive outcomes suggest that bias of
some sort is failing to be identified in individual studies, even when conflicts of
interest are disclosed.[31]

A fourth stated justification for disclosure policies is that they help prevent
the appearance of bias. Transparency allows human subjects and the general
public to remain confident in the integrity of science. It permits them to know
when financial conflicts are present and thus lets them come to their own conclusions about whether those interests are problematic. There are however several problems with this argument. First, as we argued above, it is not clear how
disclosing conflicts of interests would assist anyone in identifying bias. Most
people have neither the training nor the time to scrutinize research in ways that
would allow them to assess the presence and role of bias. If subjects or the public are unable to judge the effects of financial conflicts in the research, then is
difficult to see how they can come with justified conclusions about whether such
interests are problematic or not.

Second, though the evidence is scarce, some studies suggest that disclosing
financial conflicts to research subjects for instance, has no significant effect on
their willingness to participate in the research in question.[32] A very small minority seem to be troubled by financial interest that result from investigators having
equity interest but they seem to be unconcerned by other sources of financial interests such as payments from sponsors to investigators.[33]

Third, while there is little data on whether or how declarations of competing
interests affect the public's perceptions, some studies have shown that readers of
scientific journals rate the validity and believability of research significantly
*lower* for papers that disclose the existence of financial conflicts than for those

same papers when it was indicated that there were no conflicts.[34] That is, disclosures of financial conflicts appear to cause readers to have less confidence in the research, regardless of whether there is evidence that those financial interests led to bias. Thus, disclosure policies may actually reinforce, rather than reduce, the appearance of bias.

# Management

Although disclosure requirements are thought to be necessary in the task of dealing with the possible ill effects of financial conflicts, they are certainly not sufficient. Hence, most research institutions have management plans to deal with conflicts of interest that may unduly influence researchers' judgments.[35] While such management plans vary across institutions, normally they involve an oversight conflicts committee that evaluates the nature of the conflict and recommends a particular action plan, such as requiring an independent investigator or committee to evaluate the research related to the conflict, asking the investigator with the conflict to reduce the financial gain, or preventing the researcher in question from participating in particular aspects of the research.[36]

The goal of management strategies is to ensure that the risk of bias, or appearance of bias, is limited or eliminated. Management strategies attempt to reduce or eliminate the risk of bias by safeguarding aspects of research that are particularly vulnerable to the (unconscious) influence of financial interests. For example, researchers with significant financial interests in the success of a drug might be prohibited from participating in recruitment for clinical trials so that they do not intentionally or unintentionally select subjects who might be likely to respond to the drugs under study in particular ways. Similarly, conflict of interest committees might require an independent investigator or committee to review research design, methodology, or statistical analyses, so as to ensure that research is free from bias.

Yet there are several reasons why the management strategies adopted by institutions may not be very effective in reducing or eliminating bias. First, it is not clear that requiring investigators with significant financial interests to refrain from participating in particular areas of their research or asking them to divest themselves of their financial interests will be particularly effective. As mentioned earlier, the recent NIH policy severely restricting investigators financial conflicts raised significant criticism and was accompanied by the resignation of several researchers.[37] Second, current management strategies can fail to be effective for many of the same reasons that disclosure policies fail. That is, they require individuals to be able to evaluate whether or not financial interests have biased methodological or other scientific judgments. But such judgments will not always be transparent to oversight committees. Investigators must decide how to frame research question, make a host of methodological decisions in conducting research, decide what to count as data for or against their hypothesis, and determine how they will report the results of the research. As we have seen, these judgments are rarely black or white. Thus, a variety of different research questions, methodologies, or characterizations of data might be seen as legitimate and it would be difficult for others to judge whether financial interests were influencing these decisions.

Third, even if management strategies were successful in eliminating bias from the research produced by investigators with significant financial conflicts, such strategies are unlikely to correct the effects that financial conflicts can have in the publication activities of investigators. For instance, oversight committees might be able to carefully scrutinize the research of an investigator who receives significant consulting fees from a particular drug company. But committees are unlikely to prevent that researcher from publishing research casting doubts on the claims of competitors, downplaying uncertainty, or minimizing the importance of particular risks.[38] For example, the tobacco industry funded many scientists to publicly criticize studies that linked smoking to a variety of health problems, solely as a way to manufacture doubt about the state of the scientific evidence and ensure that no one would perceive scientific consensus.[39]

Fourth, management policies adopted by universities assume that the individuals responsible for evaluating and managing conflicts are themselves "independent" or "disinterested" with respect to the financial interests at stake. Yet this is unlikely to be the case. For example, individual administrators who serve on such committees will often have indirect interests in the success of investigators with financial conflicts. Although federal sources, particularly the NIH, remain the largest contributor to academic biomedical research expenditures, industry provides a considerable amount of funding.[40] Thus, administrators within universities have significant interests in maintaining and increasing their institution's relationships with industry so that they can continue funding faculty, equipment, travel, and research assistants. In addition, they are aware that universities can benefit from patents of new drugs or technologies that result from academic research. Administrators also know that universities enhance their status and reputations as research institutions through academic-industry partnerships so that they are better positioned to be successful in obtaining both private and public grants in the future.

Similarly, other individuals who would be tasked with implementing management plans in particular research contexts are likely to have significant interests in the success of the research. For example, other faculty, postdocs, or graduate students who might be given responsibility for human subject recruitment or "independent" statistical analysis when the principal investigator has a conflict of interest may have interests in the funding that results for their department, the continued respect of their colleague, or the additional prestige for their department that results from the industry partnership.

Hence oversight committees are normally comprised of administrators or other faculty who share an institutional interest in the success of the research or in the ability of the institution to continue to receive private funding. Thus, even if such the members of the committee have no direct financial conflicts in relation to the research they are overseeing, their assessments about whether research protocols or methods are appropriate, or whether statistical analyses are misleading, could also be influenced by unconscious bias.

# Prohibition or Elimination

The management strategies discussed involve the possibility of prohibiting a researcher from participating in aspects of the research where she has financial stakes. But some institutions also have policies that prohibit outright particular kinds of financial relationships because they are thought to be excessively fraught with risks of bias.[41] Some of these prohibitions involve a "rebuttable presumption" that allows for exceptions.[42] In these cases, there is a presumption that investigators will be prohibited from conducting research involving human participants when they have a financial stake in its outcome, unless they can show that there is some compelling reason why an exception must be made.[43] The prohibition thus is not absolute and can be rebutted if compelling circumstances make the involvement of the researcher in question justified.[44]

Clearly the goal of prohibiting or eliminating particular types of conflicts of interest is to prevent possible bias in the conduct and reporting of scientific research. Yet current policies that prohibit or eliminate conflicts of interest also suffer from a variety of problems and thus are unlikely to ultimate achieve the goal of preventing research bias.

First, usually the prohibition only applies to individuals who are conducting human subjects research. Clearly, there might be good reasons to prevent investigators with a significant financial interest in human subjects research from conducting that research. Where subjects are involved, bias can affect not just the objectivity of the research, but also the safety of the participants. Yet basic science research that does not directly involve human subjects can also be subsequently relied upon when proceeding to clinical trials. Thus, if bias is present in basic research, both the integrity of such research and the safety of future subjects might still be at risk.

Moreover, usually this prohibition applies only to phase I trials, which is insufficient to protect the safety of subjects and the integrity of research in phase II and phase III clinical trials. And of course, if biased research occurs in phase II or III clinical trials, it can put at risk not just subjects participating in research studies but also future patients. Consider, for example, the case of antidepressants for children, where the pharmaceutical industry failed to report the increased suicide risk in children taking some selective serotonin reuptake inhibitor antidepressants. Similarly, in the case of rofecoxib, inappropriate analysis of data concealed an increased cardiovascular risk in adults taking the drug. These cases show that biased research has significant effects on the health of future patients and not just those who enroll in phase I trials.[45]

Second, current prohibitions are too weak in that they allow for exceptions that are typically determined by oversight committees. Committees are usually asked to consider the nature of the research, the degree of risk that the research protocol poses for its human subjects, the unique qualifications of the researcher with the conflict to perform the study, and the degree to which the financial interest is related to the research. But, as we have already argued, answers to many of these and similar questions might be influenced by institutional commitments bias and thus it is unclear how effective this police can be in eliminating bias.

# Understanding Bias

We have argued that current conflict of interest policies fail to promote their aims of preventing bias, identifying bias, or reducing the appearance of bias for a variety of reasons. Our analysis reveals, however, that part of the problem may be that such policies have been informed by a problematic understanding of bias and an incorrect view of how bias operates in scientific research. First, conflict of interest policies appear to be grounded on the assumption that bias is only likely to occur in individual researchers with direct financial interests in the research. Yet as we have argued, individuals tasked with enforcing and carrying out conflict of interest policies can also have indirect or institutional interests in the success of the research.

A second assumption underlying current conflict of interest policies is that bias can be easily identified by simply scrutinizing research more closely in order to ensure that scientists adhere to scientific norms, methods, and practices. While this might be true in cases of gross scientific misconduct, researchers must make judgments throughout the research process that can be influenced by a variety of legitimate and illegitimate non-epistemic factors. Some research judgments will simply not be transparent to those who are evaluating the research for bias. And even when such judgments are transparent, it will not always be easy to determine whether financial interests have affected those judgments.

For example, a breast cancer drug trial may be designed to test whether the drug improves survival rates or it may be designed to test optimal treatment duration. From a profit motive, however, there is little incentive to test optimal treatment duration if it would decrease the amount of time that patients would take the drug. Similarly, researchers can design trials to determine whether the drug is effective when compared to placebo or to another commercialized drug. Comparative trials might allow a company to claim that their drug is as effective as a competitor when neither is much more effective than a placebo. How the research question is framed can have very different economic implications. Scientists with a financial interest in the success of the drug under investigation might allow such interests to influence the questions they ask but determining that such has been the case is not a matter of simply scrutinizing research more carefully.

Likewise, methodological choices are not always straightforward. Even in randomized controlled trials, decisions need to be made about eligibility criteria, randomization methods, concealment of allocation methods, degree of masking, and analytic approach. And in designing clinical trials, researchers must make decisions about what sorts of side effects to look for, including what counts as a potentially "harmful" side effect, the duration of follow-up, as well as the selection of end points, and the comparison group. Two different methodological options might both have some reasons supporting them so that even if financial interests contributed to judgments of adopting one over the other, it will be diffi-cult to identify the extent to which the selection of methods was biased.

But adopting one methodology over another could favor the safety or effi-cacy of the intervention being tested, and thus could have important economic consequences. For instance, researchers testing an antidepressant could choose whether or not to inquire about particular side effects such as sexual problems. This choice could make a preferred antidepressant appear safer than otherwise or could emphasize a competitor's side effects.[46] Or, in testing the efficacy of cardiovascular interventions, researchers' choice to use surrogate outcome measures such as quantitative angiography, intravascular ultrasound, plasma bi-omarkers, or clinical end points could affect trial results.[47] Thus, the assumption that researchers are impartial to the extent that they adhere to scientific norms does not appear to be sufficient for avoiding bias. Such norms are insufficient to guarantee that financial interests will not influence the research.

Similarly, scientists must decide what to count as data for or against their hypothesis and how to interpret their data. As in the case of research questions, and methodological decisions, determinations about how to interpret data are not always straightforward. For example, in the 1970s and 80s studies were con-ducted on the possible carcinogenic effects of dioxin. One of the most extensive studies on the toxicity of dioxins was the Kociba rat study. Over a ten-year peri-od, three groups of pathologists reviewed the rat liver slides that Kociba's team produced in order to calculate cancer rates in rats dosed with dioxin and a con-trol group.[48] In some cases, pathologists disagreed about whether particular samples were cancerous tumors. Some pathologists considered some slides to show "borderline cases" of malignancy. Others argued that the suspicious areas were slide imperfections rather than malignancies. Disagreements persisted even when new standards for identifying malignant tumors were adopted.[49] This sort of judgment can be influenced by financial interests. Indeed, of the three groups of pathologists who evaluated the Kociba rat liver slides, the group funded by the paper industry, one of the largest producers of dioxin emissions, identified the fewest number of malignant tumors. Those funded by the EPA, on the other hand, founded the greatest number.[50] Clearly, simply scrutinizing research to en-sure that scientists adhere to scientific norms, methods, and practices is unlikely to determine whether the research is biased or not.

# Implications for Dealing with Conflicts of Interest

If our analysis here is correct, then attending to how bias can enter scientific re-search is essential to conflict of interest policies. If such policies are to be effec-tive, they must account for the fact that financial interests have the potential to bias a variety of scientific judgments in ways that will not always be apparent. As we have seen, scientists need to make a variety of judgments about the fram-ing of research questions, the choice of methodology, and the interpretation of data. They also need to make judgments about when to report data and what to publish. Unless conflict of interest policies attend to the ways in which these judgments are made, how they operate, when they are legitimate or illegitimate, it is unlikely that these policies will be able to protect the integrity of scientific research.

But conflict of interest policies must also take into account the fact that, except in very particular cases, it is very difficult for individuals to identify when biases have occurred in their own research or in the research of others. Bias can occur consciously and unconsciously. Conscious bias occurs, for example, in cases where researchers' personal interests lead them to deliberately fabricate data or discard evidence that undermines a favored result. Presumably, however, this occurs only in a small number of cases. And although clearly the particular individual could identify the bias, as we have indicated others might have difficulties determining whether a particular scientific judgment is the result of illegitimate concerns. This explains, at least in part, why it is difficult to identify cases of misconduct even when such research is presumably rigorously scrutinized by colleagues, reviewers, and editors.[51]

But if intentional biases are difficult to identify, more difficult still is to account for the possibility that conscientious scientists with professional integrity might *unconsciously* allow their financial interests to affect scientific reasoning. In these cases, obviously the particular researchers will be unable to identify how their financial interest might bias the way they frame research questions, or how they interpret the data, or how they design a clinical trial.[52] And because the subtlety of these influences others might have similar difficulties determining whether the research is biased.

No less important for the drafting of effective conflict of interest policies is the need to take into account the fact that institutions also have financial interests that can inappropriately influence the judgments of individuals tasked with identifying, managing, and overseeing financial conflict of interest.[53] These requirements for academic institutions and federal agencies to reflect on funding issues, commercial relationships, the role of academic science in generating knowledge and improving public well-being, the relevance of traditional academic values and how commercial interest might affect them.

Given these considerations, what then can academic institutions and federal agencies do in order to ensure that financial interest will not affect the integrity of scientific research? One obvious solution would be to prohibit scientists and academic institutions from receiving any type of financial support from companies that could benefit from the results of the research, or preventing them from having any financial stakes on the research they do. Similarly, journals could refuse to publish studies where any financial conflicts of interest exist.

However, although these measures might indeed eliminate the existence of financial conflicts, given decreasing federal resources for research and the increasing emphasis on academic-industry partnerships, such proposal seems unrealistic. Moreover, if financial conflicts of interest do not necessarily lead to bias, then to prohibit financial relationships between scientists and private companies might negatively affect scientific and technological innovation as well and public support since these relationships have contributed to improving public health.[54]

A more feasible alternative to current conflict of interest policies would be to limit contractual provisions currently accepted by universities and academic researchers when working with industry.[55] For example, alternative policies might ensure that investigators have full participation in the methodology and design of clinical trials. Academic institutions should ensure that industry contracts allow researchers to have complete access to data and methodological decisions. Sponsors should not be allowed to add their own statistical analyses to final manuscripts, terminate trials for commercial reasons, or suppress results they find unfavorable.[56] Similarly, requirements to register clinical trials can contribute to ensuring that unfavorable trial results are made public, taken into account, and scrutinized and drug-related risks are clearly reported.[57]

In addition, while it may be difficult for individuals to identify bias in themselves or others, increasing avenues for the critical evaluation of research may enable the relevant community of experts to provide a system of checks and balances in helping to prevent the effects of unconscious bias. Greater transparency of data and methodological choices would open up more avenues for critical evaluation by reviewers and other peer researchers. For example, some journals offer now the option to researchers to agree to report more detailed descriptions of trial design, methodological choices, and data interpretation in online versions of their articles. New efforts to create resources to develop and maintain up-to-date information, tools, and other materials to help improve the quality of reporting health research can also contribute to ensure that complete and clear research information is available for scrutiny and that consistent guidelines are used in reporting of health related research.[58]

Communities of researchers could also help prevent bias by putting more resources into attempting to replicate results of successful research. Replication studies however, are not very highly valued as they are not viewed as "original" scientific work, and are often difficult to perform.[59] Such studies, however, are one reliable tool for verifying the quality and validity of original research.[60]

# Conclusion

If information about positive outcomes reporting in studies funded by industry are any indication, concerns about the problematic influence of financial interest in scientific research are certainly sensible. However, the current emphasis on conflict of interest policies is likely to be insufficient in preventing such damaging influences.

As we have shown, although the main goal of these policies is purportedly to safeguard the objectivity of scientific research, they fail to do so. In part, this is so, because conflict of interest policies presuppose a problematic understanding of how bias can affect research and thus they are likely to use ineffective mechanisms in trying to prevent such bias. Rather than concentrating all efforts to prevent bias in drafting, implementing and enforcing these policies, academic institutions, federal agencies, and researchers might better protect scientific objectivity by also focusing on other types of measures, such as limiting problematic contractual provisions in research sponsored by industry, ensuring that replication of studies occurs, or developing resources that allow for clear and transparent reporting of all relevant research information. Unlike conflict of interest policies, these measures help safeguard the sorts of decisions—about which questions to ask, which methodologies to employ, how to interpret data, what data to publish—that can be improperly influenced by financial interests.

# Notes

1. See Kaiser, J. "Private money, public disclosure." *Science* 325 (2009): 28–30; Brody, B. "Intellectual property and biotechnology: the U.S. internal experience-Part I." *Kennedy Inst Ethics J.* 16, no. 1(2006): 1-37; Mowery, D., R. Nelson, B. Sampat, and A. Ziedonis. *Ivory tower and industrial innovation—University-industry technology transfer before and after the Bayh–Dole Act.* Stanford: Stanford Business Books, 2004.

2. Sterckx, S. "Patenting and Licensing of University Research: Promoting Innovation or Undermining Academic Values?" *Science and Engineering Ethics* (Sep 19, 2009).

3. Loewenberg, S. "The Bayh-Dole Act: a model for promoting research translation?" *Mol Oncol.* 3, vol. 2 (2009): 91-3.

4. Dorsey E.R., J. de Roulet, J.P. Thompson, J.I. Reminick, A. Thai, Z. White-Stellato, C.A. Beck, B.P. George, and H. Moses 3rd. "Funding of US biomedical research, 2003-2008." *JAMA* 3032 (2010): 137-43.

5. See Geiger, R., & C. Sa. *Tapping the riches of science. Universities and the promise of economic growth.* Cambridge, MA: Harvard University Press, 2008; So, A.D., B.N. Sampat, A.K. Rai, R. Cook-Deegan, J.H. Reichman, R. Weissman, and A. Kapczynski. "Is Bayh-Dole good for developing countries? Lessons from the US experience." *PLoS Biol.* 6, no. 10 (Oct 28, 2008): e262; Boettiger, S., A.B. Bennett, and Bayh-Dole. "If we knew then what we know now." *Nat Biotechnol.*24, no. 3 (March 2006): 320-3; Campbell, E.G., B.R. Clarridge, M. Gokhale, et al. "Data withholding in academic genetics: evidence from a national survey." *JAMA* 287 (2002): 473-80.

6. See Nkansah, N., T. Nguyen, H. Iraninezhad, and L. Bero. "Randomized trials assessing calcium supplementation in healthy children: relationship between industry sponsorship and study outcomes." *Public Health Nutr.*12, no. 10 (2009): 1931-7; Khan S.N., M.J. Mermer, E. Myers, and H.S. Sandhu. "The roles of funding source, clinical trial outcome, and quality of reporting in orthopedic surgery literature." *American Journal of Orthopedics (Belle Mead NJ)* 37, no. 12 (Dec 2008): E205-12; Sismondo, S. "Pharmaceutical company funding and its consequences: a qualitative systematic review." *Contemporary Clinical Trials* 29, no. 2 (2008): 109–113; Ridker, P.M. and J. Torres. "Reported outcomes in major cardiovascular clinical trials funded by for-profit and not-for-profit organizations: 2000-2005." *JAMA* 295, no. 19 (2006): 2270-4; Bhandari, M., J.W. Buss, D. Jackowski, V.M. Montori, H. Schünemann, S. Sprague, D. Mears, E.H. Schemitsch, D. Heels-Ansdell, and P.J. Devereaux. "Association between industry funding and statistically significant pro-industry findings in medical and surgical randomized trials." *CMAJ* 170, no. 4 (Feb 17, 2004): 477-80.

7. See World Association of Medical Editors (WAME). "Conflict of interest in peer-reviewed medical journals: a policy statement of the World Association of Medical Editors (WAME)". *J Child Neurol.* 24, no. 10 (2009): 1321-3; Institute of Medicine (IOM). *Conflict of interest in medical research, education, and practice.* Washington, DC: National Academies Press, 2009; Association of American Medical Colleges (AAMC). *Task Force on Financial Conflicts of Interest in Clinical Research. Protecting Subjects, Pre-*

*serving Trust, Promoting Progress. Guidelines for Dealing with Faculty Conflicts of Commitment and Conflicts of Interest in Research.* Washington, DC: AAMC, 2001; Association of American Universities (AAU). *Task Force on Research Accountability. Report on Individual and Institutional Financial Conflict of Interest.* Washington, DC: AAU, 2001.

8. Association of American Medical Colleges (AAMC), *Task Force on Financial Conflicts of Interest in Clinical Research. Protecting Subjects, Preserving Trust, Promoting Progress. Guidelines for Dealing with Faculty Conflicts of Commitment and Conflicts of Interest in Research.* Washington, DC: AAMC, 2001; Association of American Universities (AAU), *Task Force on Research Accountability. Report on Individual and Institutional Financial Conflict of Interest.* Washington, DC: AAU, 2001; Institute of Medicine (IOM). *Conflict of interest in medical research, education, and practice.* Washington, DC: National Academies Press, 2009.

9. See de Melo-Martín and Intemann. "How do disclosure policies fail? Let us count the ways." FASEB J. 23, no. 6 (2009): 1638-42; Elliott K.C. "Scientific judgment and the limits of conflict-of-interest policies." *Account Res.* 15 no. 1 (2008): 1-29.

10. The relationships between pharmaceutical companies and physicians and the significant contributions this industry makes to continuing medical education have also been a significant source of conflict of interest. However, our concern here is only with conflicts of interest in biomedical research.

11. See Caplan, A. "Halfway there: the struggle to manage conflicts of interest." *J. Clin. Invest.* 117 (2007): 509-510; Horton R. "Conflicts of interest in clinical research: opprobrium or obsession?" *Lancet* 349 (1997): 1112-3.

12. Elliott K.C. "Scientific judgment and the limits of conflict-of-interest policies." *Account Res.* 15 no. 1 (2008): 16-17.

13. See Krimsky, S. *Science in the Private Interest.* Lanham, MD: Rowman and Littlefield, 2003; Elliott K.C. "Scientific judgment and the limits of conflict-of-interest policies." *Account Res.* 15 no. 1 (2008): 1-29.

14. See Rothman, K.J. "The Ethics of Research Sponsorship." *Journal of Clinical Epidemiology* 44 (1991): 25S-28S; Rothman, K.J. "Conflict of interest: the new McCarthyism in science." *JAMA* 269 (1993): 2782-2784.

15. See Krimsky, S. and L. Rothenberg. "Conflict of interest policies in science and medical journals: Editorial practices and author disclosure." *Science and Engineering Ethics* 7 (2001): 205–218; Krimsky, S. *Science in the Private Interest.* Lanham, MD: Rowman and Littlefield, 2003; Elliott K.C. "Scientific judgment and the limits of conflict-of-interest policies." *Account Res.* 15 no. 1 (2008): 1-29.

16. de Melo-Martín, I. and K. Intemann. "How do disclosure policies fail? Let us count the ways." *FASEB J.* 23, no. 6 (2009): 1638-42.

17. Institute of Medicine (IOM). *Conflict of interest in medical research, education, and practice.* Washington, DC: National Academies Press, 2009.

18. See Institute of Medicine (IOM). *Conflict of interest in medical research, education, and practice.* Washington, DC: National Academies Press, 2009; AMSA. *PharmFree Scorecard 2008.* Reston, VA: AMSA, 2008. http://amsascorecard.org/ (accessed July 10, 2010).

19. Association of American Medical Colleges (AAMC). *Task Force on Financial Conflicts of Interest in Clinical Research. Protecting Subjects, Preserving Trust, Promoting Progress. Guidelines for Dealing with Faculty Conflicts of Commitment and Conflicts of Interest in Research.* Washington, DC: AAMC, 2001; Association of American Universities (AAU). *Task Force on Research Accountability. Report on Individual and Institutional Financial Conflict of Interest.* Washington, DC: AAU, 2001; Institute of Medi-

cine (IOM). *Conflict of interest in medical research, education, and practice*. Washington, DC: National Academies Press, 2009.

20. Ehringhaus S. and D. Korn. "U.S. Medical School Policies on Individual Financial Conflicts of Interest." Results of an AAMC Survey. Washington, DC: AAMC, 2004.

21. See AMSA. *PharmFree Scorecard 2008*. Reston, VA: AMSA, 2008. http://amsascorecard.org/ (accessed July 10, 2010); Weinfurt, K.P., M.A. Dinan, J.S. Allsbrook, J.Y. Friedman, M.A. Hall, K.A. Schulman, and J. Sugarman. "Policies of academic medical centers for disclosing financial conflicts of interest to potential research participants." *Academic Medicine*. 81, no. 2 (2006): 113–118; Ehringhaus S. and D. Korn. "U.S. Medical School Policies on Individual Financial Conflicts of Interest." Results of an AAMC Survey. Washington, DC: AAMC, 2004.

22. See Institute of Medicine (IOM). *"Conflict of interest in medical research, education, and practice."* Washington, DC: National Academies Press, 2009; Drazen and Curfman. "Financial associations of authors." *New England Journal of Medicine* 346 (2002):1901-2; Campbell, Philip. "Introducing a new policy for authors of research papers in *Nature* and *Nature* journals." *Nature* 412, 751 (August 23 2001): 751; DeAngelis, C.D., P.B. Fontanarosa, and A. Flanagin. "Reporting financial conflicts of interest and relationships between investigators and research sponsors." *JAMA* 286 (2001): 89-91.

23. See Campbell, E.G. "Doctors and drug companies—scrutinizing influential relationships." *New England Journal of Medicine* 357, no. 18 (2007): 1796-7; Boyd, E.A., M.K. Cho, and L.A. Bero. "Financial conflict-of-interest policies in clinical research: issues for clinical investigators." *Acad Med* 78 (2003): 769-74; Charatan, F. "Doctors say they are not influenced by drug companies' promotions." *BMJ* 322 (2001): 1081.

24. See Smith, C.D. and B. MacFadyen. "Industry relationships between physicians and professional medical associations: corrupt or essential?" *Surg Endosc.* 24, no. 2 (Feb, 2010): 251-3; Hirsch L.J. "Conflicts of interest, authorship, and disclosures in industry-related scientific publications: the tort bar and editorial oversight in medical journals." *Mayo Clinic Proceedings* 84, no. 9 (2009): 811-821; Stossel, T.P. "Regulating academic industry research relationships—solving problems or stifling progress?" *New England Journal of Medicine* 353, no. 10 (2005): 1060-1065; Lipton, S., E.A. Boyd, and L.A. Bero. "Conflicts of interest in academic research: policies, processes, and attitudes." *Account Res.* 11 (2004): 83-102.

25. See Wadman, M. "NIH workers see red over revised rules for conflicts of interest." *Nature* 434 (2005): 3-4; Dalton, R. "Postdocs slam zealous attitude of NIH ethics office." *Nature* 434 (2005): 687.

26. See Kaiser, J. "Conflict of interest. NIH rules rile scientists, survey finds." *Science* 314 (2006): 740; Twombly, R. "Conflict-of-interest rules worry some scientists." *J Natl Cancer Inst.* 99 (2007): 6-9.

27. Boyd, E.A., M.K. Cho, and L.A. Bero. "Financial conflict-of-interest policies in clinical research: issues for clinical investigators." *Acad Med* 78 (2003): 769-74.

28. See Cain, Daylian M. and Allan S. Detsky. "Everyone's a Little Bit Biased (Even Physicians)." *JAMA.* 299, no. 24 (2008): 2893-2895; Katz D., A.L. Caplan, and J.F. Merz. "All gifts large and small: toward an understanding of the ethics of pharmaceutical industry gift-giving." *American Journal of Bioethics* 3, no. 3 (2003): 39-46.

29. Cain, Daylian M., G. Loewenstein, and D.A. Moore. "The dirt on coming clean: perverse effects of disclosing conflicts of interest." *J Legal Stud.* 34 (2005): 1-25.

30. de Melo-Martín, I. and K. Intemann. "How do disclosure policies fail? Let us count the ways." *FASEB J.* 23, no. 6 (2009): 1638-42.

31. See Nkansah, N., T. Nguyen, H. Iraninezhad, and L. Bero. "Randomized trials assessing calcium supplementation in healthy children: relationship between industry sponsorship and study outcomes." *Public Health Nutr.*12, no. 10 (2009): 1931-7; Sis-

mondo, S. "Pharmaceutical company funding and its consequences: a qualitative systematic review." *Contemporary Clinical Trials* 29, no. 2 (2008): 109–113; Ridker, P.M. and J. Torres. "Reported outcomes in major cardiovascular clinical trials funded by for-profit and not-for-profit organizations: 2000-2005." *JAMA* 295, no. 19 (2006): 2270-4; Bhandari, M., J.W. Buss, D. Jackowski , V.M. Montori, H. Schünemann, S. Sprague, D. Mears , E.H. Schemitsch, D. Heels-Ansdell, and P.J. Devereaux. "Association between industry funding and statistically significant pro-industry findings in medical and surgical randomized trials." *CMAJ* 170, no. 4 (Feb 17, 2004): 477-80.

32. Weinfurt, K.P., M.A. Hall, M.A. Dinan, V. DePuy, J.Y. Friedman, J.S. Allsbrook, and J. Sugarman. "Effects of disclosing financial interests on attitudes toward clinical research." *J Gen Intern Med.* 23, no. 6 (2008): 860-6.

33. Weinfurt, K.P., M.A. Hall, M.A. Dinan, V. DePuy, J.Y. Friedman, J.S. Allsbrook, and J. Sugarman. "Effects of disclosing financial interests on attitudes toward clinical research." *J Gen Intern Med.* 23, no. 6 (2008): 860-6.

34. Schroter, Morris, Chaudhry, Smith, and Barratt. "Does the type of competing interest statement affect readers' perceptions of the credibility of research? Randomised trial." *BMJ* 328 (2004): 742-3; Editorial. "Dealing with disclosure." *Nat Med* 12 (2006): 979.

35. See AMSA. *PharmFree Scorecard 2008.* Reston, VA: AMSA, 2008. http://amsascorecard.org/ (accessed July 10, 2010); Ehringhaus S. and D. Korn. "U.S. Medical School Policies on Individual Financial Conflicts of Interest." Results of an AAMC Survey. Washington, DC: AAMC, 2004.

36. See Institute of Medicine (IOM). *Conflict of interest in medical research, education, and practice.* Washington, DC: National Academies Press, 2009; Lipton, Boyd, and Bero. "Conflicts of interest in academic research: policies, processes, and attitudes." *Account Res.* 11 (2004): 83-102; Ehringhaus and Korn. "U.S. Medical School Policies on Individual Financial Conflicts of Interest." Results of an AAMC Survey. Washington, DC: AAMC, 2004.

37. See Wadman, M. "NIH workers see red over revised rules for conflicts of interest." *Nature* 434 (2005):3-4; Dalton, R. "Postdocs slam zealous attitude of NIH ethics office." *Nature* 434 (2005): 687.

38. Elliott K.C. "Scientific judgment and the limits of conflict-of-interest policies." *Account Res.* 15 no. 1 (2008): 1-29.

39. Michaels, D. *Doubt is their product.* New York: Oxford University Press, 2008.

40. Dorsey E.R., J. de Roulet, J.P. Thompson, J.I. Reminick, A. Thai, Z. White-Stellato, C.A. Beck, B.P. George, and H. Moses 3rd. "Funding of US biomedical research, 2003-2008." *JAMA* 3032 (2010): 137-43.

41. Institute of Medicine (IOM). *Conflict of interest in medical research, education, and practice.* Washington, DC: National Academies Press, 2009.

42. Ehringhaus S. and D. Korn. "U.S. Medical School Policies on Individual Financial Conflicts of Interest." Results of an AAMC Survey. Washington, DC: AAMC, 2004.

43. Association of American Medical Colleges (AAMC). *Task Force on Financial Conflicts of Interest in Clinical Research. Protecting Subjects, Preserving Trust, Promoting Progress. Guidelines for Dealing with Faculty Conflicts of Commitment and Conflicts of Interest in Research.* Washington, DC: AAMC, 2001.

44. Association of American Medical Colleges (AAMC). *Task Force.*

45. See Lagakos, S.W. "Time-to-event analyses for long-term treatments: the APPROVe trial." *New England Journal of Medicine* 355 (2006):113-117; Curfman, G.D., S. Morrissey, and J.M. Drazen. "Expression of concern." *New England Journal of Medicine* 343 (2000): 1520-8; Whittington, C.J., T. Kendall, P. Fonagy, D. Cottrell, A. Cotgrove,

and E. Boddington. "Selective serotonin receptor inhibitors in childhood depression: systematic review of published versus unpublished data." *Lancet* 363 (2004): 1341–1345.

46. Safer D.J. "Design and reporting modifications in industry-sponsored comparative psychopharmacology trials." *J Nerv Ment Dis.* 190, no. 9 (2002): 583–592.

47. Ridker, P.M. and J. Torres. "Reported outcomes in major cardiovascular clinical trials funded by for-profit and not-for-profit organizations: 2000-2005." *JAMA* 295, no. 19 (2006): 2270-4.

48. Douglas, H. "Inductive risk and values in science." *Philosophy of Science* 67, no. 4 (2000): 559-579.

49. Douglas, H. "Inductive risk," 559-579.

50. Douglas, H. "Inductive risk," 559-579.

51. See Rusnak and Chudley. "Stem cell research: cloning, therapy and scientific fraud." *Clin Genet.* 70, no. 4 (Oct, 2006): 302-5; Normile, Vogel, and Couzin. "Cloning. South Korean team's remaining human stem cell claim demolished." *Science* 311, no. 5758 (2006): 156-7; Horton, "Expression of concern: non-steroidal anti-inflammatory drugs and the risk of oral cancer." *Lancet* 367, no. 9506 (2006): 196.

52. Cain, Daylian M. and Allan S. Detsky. "Everyone's a Little Bit Biased (Even Physicians)." *JAMA.* 299, no. 24 (2008): 2893-2895.

53. Slaughter, S., M.P. Feldman, and S.L. Thomas. "U.S. research universities' institutional conflict of interest policies." *J Empir Res Hum Res Ethics* 4, no. 3 (2009): 3-20.

54. See Schnittker, J. and G. Karandinos. "Methuselah's medicine: pharmaceutical innovation and mortality in the United States, 1960-2000." *Soc Sci Med.* 70, no. 7 (2010): 961-8; Smith, C.D. and B. MacFadyen. "Industry relationships between physicians and professional medical associations: corrupt or essential?" *Surg Endosc.* 24, no. 2 (Feb, 2010): 251-3; Seltzer S.E., A. Menard, R. Cruea, and R. Arenson. "'Hyperscrutiny' of academic-industrial relationships: potential for unintended consequences—a response." *Journal of the American College if Radiol.* 7, no. 1 (2010): 39-42; Zycher, B., J.A. DiMasi, and C.P. Milne. "Private sector contributions to pharmaceutical science: thirty-five summary case histories." *Am J Ther.*17, no. 1 (2010): 101-20.

55. See Mello, M.M., B.R. Clarridge, and D.M. Studdert. "Academic medical centers' standards for clinical-trial agreements with industry." *New England Journal of Medicine* 352, no. 21 (2005): 2202-2210; Mello, M.M., B.R. Clarridge, and D.M. Studdert. "Researchers' views of the acceptability of restrictive provisions in clinical trial agreements with industry sponsors." *Account. Res.* 12, no. 3(2005): 63-91.

56. See Mello, Clarridge, and Studdert. "Academic medical," 2202-2210; Mello, Clarridge, and Studdert. "Researchers' views of the acceptability of restrictive provisions in clinical trial agreements with industry sponsors." *Account. Res.* 12, no. 3(2005): 63-91.

57. See Abaid, Grimes, and Schulz. "Reducing publication bias through trial registration." *Obstet Gynecol* 109 (2007): 1434-7; Laine, Horton, DeAngelis, Drazen, Frizelle, Godlee, Haug, Hébert, Kotzin, Marusic, Sahni, Schroeder, Sox, Van der Weyden, and Verheugt. "Clinical trial registration: looking back and moving ahead." *JAMA.* 298, no. 1 (2007): 93-4.

58. See Simera, I., D. Moher, A. Hirst, J. Hoey, K.F. Schulz, and D.G. Altman. "Transparent and accurate reporting increases reliability, utility, and impact of your research: reporting guidelines and the EQUATOR Network." *BMC Med.* 8 (2010): 24; Altman D. G., I. Simera, J. Hoey, D. Moher, and K. Schulz. "EQUATOR: reporting guidelines for health research." *Lancet* 371 (2008): 1149–1150.

59. van den Oord, EJCG. "Controlling false discoveries in genetic studies." *Am J Med Genet B Neuropsychiatr Genet.*147B (2008): 637–44.

60. See Ioannidis, J., E. Ntzani, T. Trikalinos, and D. Contopoulos-Ioannidis. "Replication validity of genetic association studies." *Nat Genet.* 29, no. 3 (Nov 2001): 306-9;

Colhoun, H.M., P.M. McKeigue, and G.D. Smith. "Problems of reporting genetic associations with complex outcomes." *Lancet* 361 (2003): 865–72.

# Chapter 13

# Science, Religion, and Duty in Parenting Choices

## Glenn Sanford

In September 2009, C.S. Mott Children's Hospital released the results of a national poll studying parental attitudes toward H1N1 influenza and the recommended vaccine.[1] Despite the Centers for Disease Control (CDC) identifying children as a priority group for vaccinations, only 40 percent of those surveyed responded "Definitely/Probably Yes" when asked whether they planned to have their children vaccinated for H1N1.[2] A series of AP-Gfk Polls taken during July, September, October, and November 2009 found that 36-38 percent of parents were unlikely to give permission to have their children vaccinated at school.[3] Despite the success of vaccination programs[4] and a 1905 Supreme Court decision upholding the constitutionality of mandatory vaccination laws[5], resistance to childhood vaccinations remains a major issue within public health circles.[6] In what follows, I will first review the civil and criminal liability of parents for their childrearing choices. Then, following a brief review of the case law concerning compulsory immunization law, I will conclude by considering the conditions under which parents may be subject to criminal penalties for refusing to vaccinate their children. Interestingly, the application of immunization laws turns more on the duties prescribed by statute than on the science of epidemiology or religious freedom. As will be discussed, in the United States, legislatures possess the power to weigh science, religion, and politics in creating mandatory immunization laws.

## Parental Liability: Civil and Criminal Actions Under State Law

As if the relationship between parents and children wasn't complex enough, a South Carolina statute allows a person, who would otherwise be subject to the death penalty for a stabbing resulting in death, to escape that fate if the stabbing occurred "in chastising or correcting his child."[7] In 1891, the Mississippi Supreme Court, concluding that a minor daughter could not sue her mother for damages sustained as a result of her wrongful confinement to an insane asylum, stated:

> [S]o long as the parent is under obligation to care for, guide, and control, and the child is under reciprocal obligation to aid and comfort and obey, no such action as this can be maintained. The peace of society, and of the families composing society, and a sound public policy, designed to subserve the repose of families and the best interests of society, forbid to the minor child a right to appear in court in the assertion of a claim to civil redress for personal injuries suffered at the hands of the parent. The state, through its criminal laws, will give the minor child protection from parental violence and wrong-doing, and this is all the child can be heard to demand.[8]

The court offered no precedent or citation for this statement of the parent-child relationship; however, as will be discussed, courts traditionally provided parents with broad immunity to suits for harming their own children.[9] In a few lines, the Mississippi court captured the legal duality of parent-child relations; namely, though a child was traditionally prevented from suing a parent for damages, the state has long held the power to criminalize a parent's behavior toward the child.[10] It is against this backdrop that I will examine the interaction of parents' rights, children's rights, religious freedom, and the power of the state to provide for the health, safety, and welfare of its citizens. A brief review of case law concerning the extent of parents' rights concerning childrearing decisions will set the stage for considering the potential issues presented when parents refuse to immunize their children.

In 1903, New York's highest court upheld the criminal conviction of a J. Luther Pierson, whose adopted daughter died without receiving medical care.[11] He did not believe in using physicians and believed that "Divine healing" would cure her.[12] In rejecting the father's claim that the application of the criminal statute requiring parents to provide medical care for children violated his First Amendment right to free exercise of religion, the court stated, "Full and free enjoyment of religious profession and worship is guaranteed, but acts which are not worship are not."[13] Though the opinion did not cite *Reynolds v. United States*, it reflects the U.S. Supreme Court's position therein, which holds:

> Laws are made for government of actions, and while they cannot interfere with mere religious beliefs and opinions, they may with practices. . . . Can a man excuse his practices to the contrary because of his religious belief? To permit this would make the professed doctrines of religious belief superior to the law of the land, and in effect to permit every citizen to become a law unto himself. Government could exist only in name under such circumstances.[14]

In the *Pierson* case, the New York court specifically avoided addressing any question with respect to the relative truth or value of science and religion; instead, it simply pointed out that this was a case of a man who had failed to do what was required of him by a statute.[15] Despite the court's stated deference with respect to the relative merits of science and religion, in the end, Mr.

Pierson's conviction under the law was upheld because:

> The legislature is the sovereign power of the state. It may enact laws for the maintenance of order by prescribing a punishment for those who transgress. While it has no power to deprive person of life, liberty, or property without due process of law, it may in the case of commission of acts which are public wrongs or which are destructive of private rights, specify that for which the punishment shall be death, imprisonment or the forfeiture of property.[16]

Thus, this opinion offers a glimpse of the rule that the legislature has extensive power to enact laws criminalizing activities that are contrary to its vision of the public welfare, so long as those prosecuted are afforded due process.

In contrast to the rather direct endorsements of legislative power discussed thus far, the Court, in *Pierce v. Society of Sisters*, held that "[R]ights guaranteed by the constitution may not be abridged by legislation which has no reasonable relation to some purpose within the competency of the state. . . . The child is not the mere creature of the state . . ."[17] Likewise, in *Prince v. Massachusetts*, the court reiterated that "It is cardinal with us that the custody, care, and nurture of the child reside first in the parents, whose primary functions and freedom include preparation for obligations the state can neither supply nor hinder."[18] In *Parham v. J.R.*, the court added, "The statist notion that governmental power should supersede parent authority in *all* cases because *some* parents abuse and neglect children is repugnant to American tradition" (emphasis in original).[19] All of this led the Tenth Circuit Court of Appeals to conclude, "the constitutional right to family integrity is amorphous and always must be balanced against the government interest involved."[20] The central holding in these and myriad other cases is that while the state possesses expansive power to regulate the public welfare, parents enjoy constitutional protections to the extent that their child rearing decisions that must be weighed against the perceived safety of the child or general welfare of society. From *Wisconsin v. Yoder*, we learn, "The essence of all that has been said and written on the subject is that only those interests of the highest order and those not otherwise served can overbalance legitimate claims to the free exercise of religion."[21] It is this balancing that is the central issue in the context of compulsory immunization laws.

# Compulsory Immunization: Exercising Police Power for the Public Good

The broad powers of the legislature to exercise its police power to promote the health, safety, and welfare of the citizens they represent have long been recognized by the Supreme Court.[22] With respect to immunizations, the Court first weighed in on the debate in 1905, when in *Jacobson v. Massachusetts*, it found that states' police power extended to the enactment of compulsory vaccination laws.[23] Applying the aforementioned doctrine of state power, the court held:

The mode or manner in which those results are to be accomplished is within the discretion of the state, subject, of course, so far as Federal power is concerned, only to the condition that no rule prescribed by a state, nor any regulation adopted by a local governmental agency acting under the sanction of state legislation, shall contravene the Constitution of the United States, nor infringe any right granted or secured by that instrument.[24]

Confronting what it sees as the potential for anarchy that would arise if individuals were possessed of the power to substitute their own conscience for the general laws, the Court emphasizes the necessity of the legislature "secur[ing] the general comfort, health, and prosperity of the state" within a context of "liberty regulated by law."[25] In this light, the court frames immunization as self-defense adding "of paramount necessity, a community has the right to protect itself against an epidemic of disease which threatens the safety of its members."[26] Having previously affirmed states' rights to criminalize the actions of parents that threaten their children's safety and well-being, in *Jacobson* the Court affirmed the legislative power to compel compliance with public health laws irrespective of conscientious objections. Applying the foregoing principles to the issue of childhood immunization, the courts have made clear that states may require compulsory vaccinations that do not include religious exemptions provided such regulations are directly tied to the public interest.[27]

In 1920, the Texas Court of Civil Appeals presaged the future of compulsory immunization jurisprudence when it upheld the mandatory vaccination requirement of the San Antonio Independent School District holding that:

There is no greater authority or responsibility placed upon a city than to establish rules and regulations to protect the health and morals of a city. This power did not suddenly come up, but it followed the growth of civilization in the wake of necessity, progress, and science, and spread of disease, and as conditions arose and necessity demanded, the inherent power to grapple with it sprang into life and action, to the end that the blighting hand of disease might not touch and destroy the inhabitants.[28]

Extending the reasoning of *Jacobson*, this stance provided a template for future decisions upholding compulsory immunization rules. As an illustration, consider the 1959 decision in *Board of Education of Mountain Lakes v. Maas* as representative of the case law addressing compulsory vaccination.[29] The children of a Christian Scientist were barred from entering a public school after she refused to have them vaccinated.[30] The mother responded to the injunction barring her children by counterclaiming that the local board of education's policy interfered with her constitutionally protected religious freedom.[31] In 1957, the Mountain Lakes district adopted a policy mandating vaccination for smallpox and diphtheria in order for children to attend school, exempting only those who were deemed medically unfit for vaccination.[32] Though the policy was adopted in accordance with a statute that allowed boards to grant exemptions based on free exercise of religious principles, the Mountain Lakes policy contained no such exemption provision.[33] The court ruled that the policy was within the power granted to the board, because though the statute allowed exceptions, it did not require them.[34] Additionally, because the board had treated everyone the same (i.e., it had not granted any religious exemptions) and was clearly acting to promote the public welfare, "the right to practice religion freely does not include liberty to expose the community or the child to communicable disease or the latter to ill health or death."[35] The courts have consistently upheld compulsory vaccinations laws citing similar reasons;[36] however, the situation becomes more complex when the laws/policies provide for religious exemptions.

Every state requires children to have immunizations to enter school; however, 48 states allow religious exemptions, and 21 offer personal belief exemptions (only Mississippi and West Virginia allow only medical exemptions).[37] West Virginia's mandatory vaccination law only exempts those students that present "a certificate from a reputable physician showing that [the immunizations are] impossible or improper."[38] Some states, such as Maryland and California, simply require parents to sign a statement signifying their religious or personal objection to the vaccination procedure.[39] In line with the already discussed precedents of *Pierson, Jacobson,* and *Maas,* these policies rely on the discretion of the legislature in determining the mode and manner of protecting the public welfare. However, such exemptions open new areas of litigation such as whether the parents' beliefs are religious[40] and whether the religious beliefs are "genuine and sincere."[41] Additionally, these questions are not static features; rather, as in the case of *McCarthy v. Ozark School District,* wherein pending litigation by parents of unimmunized children that did not belong to recognized religious groups was rendered moot by a legislative expansion of the exemption provision to include personal, philosophical objections.[42] The collision of personal and religious freedom with legislative and judicial conceptions of public and child welfare will continue to generate controversies. We continue to struggle with cases in which parents refuse to get treatment for their children for religious reasons[43] or simply as a matter of parental choice.[44] Because of this, the conditions under which parents are subject to criminal prosecution for refusal or failure to vaccinate their children deserve closer scrutiny. To focus this discussion, the remainder of the paper will focus on when does a parent's vision of "best interest of the child" diverge from the state's sufficiently to merit a pediatrician referring a vaccination refusal for investigation and/or prosecution?

# Analyzing the Parents' Refusal to Vaccinate: The Line Between Religious Liberty and Criminality

The tension between the rights of parents to control the upbringing of their children and the state's interest in protecting the welfare of those children plays out in the offices of pediatricians around the country. Rather than attempting to consider the implications of various state statutes, I will consider general cases that may serve as the basis for a state agency to act against the wishes of a parent. This will place the following discussion squarely in the context of requiring an application of the balancing test weighing the state's interest in public health and child welfare against a parent's protected interests in raising her child and enjoying religious freedom. Two general categories of state concern come to mind: 1) the general population may be forced to bear the cost of an unimmunized person; and 2) the danger that an unimmunized child could serve as a vector to transmit a disease to others. While these can be subdivided, in each circumstance, the question will be, "Does the state's interest overwhelm the religious freedom of the parent?"

When parents refuse to immunize their children, there is more at stake than the cost of caring for the unimmunized person should they contract a preventable disease. The CDC acknowledges that generally prescribed vaccines carry the risk of side effects ranging from a rash or fever to seizures, brain damage, and life-threatening allergic reactions.[45] Parents who refuse to immunize their children are free-riders enjoying the benefits of having their children being protected by living in a population with high levels of immunization without bearing the costs associated with vaccination risks. From a public health perspective, immunizations work better as more people receive them; epidemiologists generally refer to this as herd immunity.[46] As more people are immunized, the number of susceptible people is decreased, and this, in turn, reduces the number of vectors by which the diseases can be spread.[47] Should this provide a pediatrician with reasonable grounds for referring a parent who refuses to vaccinate a child to the authorities for possible intervention? Though the free-rider problem presents interesting questions of justice and fairness, the precedents in *Jacboson, Pierson,* and *Maas* place this case squarely within the purview of legislative discretion.

Beyond the risks of vaccination, individuals lacking immunity present a threat to at least three groups of people by increasing the number of vectors for disease transmission. The groups are: 1) individuals who are unvaccinated under the assumed religious exemption; 2) individuals who were not vaccinated for medical reasons; and 3) individuals who were vaccinated, but failed to develop or subsequently lost immunity. From the perspective of disease vectors, these three groups are indistinguishable; however, each presents a slightly different matrix of ethical and social responsibility concerns. Individuals who were vaccinated, but failed to develop immunity, have directly taken the requisite steps to protect themselves, and individuals who are medically unable to receive vaccinations are in a position to argue that they have taken all reasonably available steps to protect themselves. However, individuals who are unvaccinated because of religious exemptions are unimmunized due to a personal choice to value their religious freedom over the potential benefits of immunization. Do any of these endanger a child to the point where a pediatrician should refer the case for investigation?

For all of the cases described in this section, science cannot force the state to intervene based upon evidence alone. As has been discussed, in our legal system, it is up to the legislature to determine the appropriate standards for the public good and child welfare.[48] Though the legislature may enact an immunization statute that lacks religious conscience or personal philosophy exemptions,[49] if it chooses to grant the exemption, there is no cause of action against those who take advantage of it. However, if we step back and take a roll call of scientific organizations, a clear pattern emerges: American Medical Association—vaccines play a vital role in securing public health;[50] American Academy of Pediatrics—"vaccines are one of the most successful medical advances of all time;"[51] the Institute of Medicine—"Vaccines are among the greatest public health accomplishments of the past century;"[52] and the American Academy for the Advancement of Science (and others)—*"Within a human rights framework, immunization is not simply a necessary medical requirement for children and a responsible public health measure; it is a right of all children, with corresponding government obligations. (emphasis added)."*[53] More directly, Omer, et al. documented increased risks of pertussis (whooping cough) infection in states with easy-to-get non-medical vaccine exemptions.[54] Another study, using CDC measles surveillance data from 1985-1992, found that those who received a vaccination exemption were on average thirty-five times more likely to contract measles than those who were vaccinated.[55] Reinforcing these findings, a third study not only confirmed the increased risks of exempted students acquiring measles and pertussis; it also found that county-level exemptor frequency was associated with the incidence rate of measles and found a statistically significant difference in the percentage of exempted students at schools with pertussis outbreaks (mean, 4.3 percent) as compared to those without outbreaks (mean, 1.5 percent; P=.001).[56] The documented risks of vaccinations notwithstanding, the foregoing is but a miniscule sample of the strong support for immunization among the scientific and medical communities; yet, the decision about whether to allow non-medical exemptions to compulsory vaccination statutes remains subject to the vagaries of political proceedings.

As should be clear, the major battleground is neither in the area of science nor in the area of constitutional religious freedom; rather, it emerges in the legislative halls where our society enacts laws. If the legislature, with access to medical-scientific testimony and presumably aware of herd immunity, makes the decision to adopt a religious or personal philosophy exemption, then it has explicitly placed higher value on religious freedom than on establishing the highest possible herd immunity and best public health outcomes. Nonetheless, in our system of jurisprudence, this victory for religious liberty is only secure until a future legislature decides that the public health threat of declining immunization rates, with concomitant increases in preventable infections, is more important than individual religious preferences.

# Notes

1. Davis, Matthew. "Parents May Underestimate the Risks of H1N1 Flu for Their Children," *CS Mott Children's Hospital National Poll on Children's Health* 8, no. 1 (September 24, 2009): 1-2.

2. Davis "Parents May Underestmate," 1-2.

3. *Ap-Gfk Polls* conducted by Gfk Roper Public Affairs & Media. Interviews conducted July 16-20, September 3-8, October 1-5, and November 5-9, 2009; http://surveys.ap.org/data%5CGfK%5CAPGfK%20Poll%20Final%20November%20FL U%20Topline%20Final%20111209.pdf (accessed August 15, 2010).

4. Maldonado, Yvonne A., "Current Controversies in Vaccination: Vaccine Safety," *Journal of the American Medical Association*, 288, no. 24 (December 25, 2002): 3155-3158.

5. *Jacobson v. Massachusetts*, 197 U.S. 11 (1905).

6. Steinhauer, Jennifer, "Public Health Risk Seen as Parents Reject Vaccines," *New York Times*, March 21, 2008, http://www.nytimes.com/2008/03/21/us/21vaccine.html (accessed August 15, 2010); Grant, Andrew, "Top 100 Stories of 2009 #1: Vaccine Phobia Becomes a Public-Health Threat," *Discover*, January 26, 2010, http://discovermagazine.com/2010/jan-feb/01 (accessed August 15, 2010).

7. Code of Laws of South Carolina 1976 §16-3-40.

8. *Hewelette v. George*, 9 So. 885, at 887 (Miss. 1891).

9. Rooney, Martin J. and Colleen M. Rooney, "Parental Tort Immunity: Spare the Liability, Spoil the Parent," *New England Law Review* 25 (1991): 1161-1184. *See* also, Hollister, "Parent-Child Immunity: A Doctrine in Search of Justification," *Fordham Law Review* 50 (1982): 489-527.

10. For examples of this, see *McKelvey v. McKelvey*, 77 S.W. 664 (Tenn. 1903) (holding that a child could not sue her father even though he insisted upon and consented to her stepmother's inhuman treatment) and *Roller v. Roller*, 79 P. 788 (Wash. 1905) (barring a child from suing her father for rape even though he had already been convicted of the crime and had been sentenced to prison). Both cases are discussed in more detail later in the paper.

11. *People v. Pierson*, 63 L.R.A. 187, 68 N.E. 243 (NY 1903).

12. *People v Pierson*, at 204.

13. *People v. Pierson*, at 211.

14. *Reynolds v. United States*, 98 U.S. 145, at 166-167 (1878).

15. *People v. Pierson*, 63 L.R.A. 187, 68 N.E. 243 (NY 1903).

16. *People v Pierson*, at 211.

17. *Pierce v. Society of Sisters*, 268 U.S. 510, at 535 (1925).

18. *Prince v. Massachusetts*, 321 U.S. 158, at 166 (1944). Nonetheless, the court did explicitly recognize the state's power to limit parental freedom in matters regarding the child's well-being.

19. *Parham v. J.R.*, 442 U.S. 584, at 603 (1979).

20. *Martinez v. Mafchir*, 35 F. 3d 1486, at 1490 (C.A.10 (N.M.), 1994).

21. *Wisconsin v. Yoder*, 406 U.S. 205, at 215 (1972).

22. *Wisconsin v. Yoder*, 406 U.S. 205, at 215 (1972); *Butcher's Union Slaughter-House & Live-Stock Landing Co., v. Crescent City Live-Stock Landing & Slaughter-House Co.*, 111 U.S. 746 (1884); *Boyd v. Alabama*, 94 U.S. 645 (1876).

23. *Jacobson v. Massachusetts*, 197 U.S. 11 (1905).

24. *Jacobson v Massachusetts* at 26.

25. *Jacobson v Massachusetts* at 26.

26. *Jacobson v Massachusetts* at 27.

27. *Davis v. State*, 451 A.2d. 107 (Md., 1982); *Wright v. DeWitt School District No. 1*, 385 S.W. 2d 644 (Ark., 1965); *Cude v. State*, 377 S.W. 2d 816 (Ark. 1965); *Board of Education of Mountain Lakes v. Maas*, 152 A.2d 394 (N.J. Super. Ct. App. App. Div., 1959); *Zucht v. King*, 255 S.W. 267 (Tex. Civ. App. 1920).

28. *Zucht v. King*, 255 S.W. 267 (Tex. Civ. App. 1920)

29. 152 A.2d 394 (N.J. Super. Ct. App. App. Div., 1959).

30. 152 A.2d 394 (N.J. Super. Ct. App. App. Div., 1959).

31. 152 A.2d 394 (N.J. Super. Ct. App. App. Div., 1959).

32. 152 A.2d 394 (N.J. Super. Ct. App. App. Div., 1959).

33. 152 A.2d 394 (N.J. Super. Ct. App. App. Div., 1959).

34. 152 A.2d 394 (N.J. Super. Ct. App. App. Div., 1959).

35. 152 A.2d 394 (N.J. Super. Ct. App. App. Div., 1959).

36. *Workman v. Mingo County Schools*, 667 F.Supp. 2d 679 (S.D.W.V. 2009); *Davis v. State*, 451 A.2d. 107 (Md., 1982); *Wright v. DeWitt School District No. 1*, 385 S.W. 2d 644 (Ark., 1965); *Cude v. State*, 377 S.W. 2d 816 (Ark. 1965); *Board of Education of Mountain Lakes v. Maas*, 152 A.2d 394 (N.J. Super. Ct. App. App. Div., 1959); *Zucht v. King*, 255 S.W. 267 (Tex. Civ. App. 1920).

37. Johns Hopkins Bloomberg School, "Vaccine Exemptions," Public Health Institute for Vaccine Safety, http://www.vaccinesafety.edu/cc-exem.htm, (accessed August 15, 2010).

38. West Virginia Code, §16-3-4.

39. Omer, Pan, Halsey, Stokley, Moulton, Navar, Pierce, Salmon, "Nonmedical exemptions to School Immunization Requirements: Secular Trends and Association of State Policies with Pertussis Incidence," *The Journal of the American Medical Association* 296 no. 14 (October 11, 2006): 1757-1763.

40. *In re Moses v. Bayport Bluepoint Union Free School District*, (not reported) 2007 WL 526610 (E.D.NY February 13, 2007); *McCarthy v. Ozark School District*, 359 F. 3d 1029 (C.A. 8 (Ark.), 2004).

41. *Caviezel v. Great Neck Public Schools*-F.Supp. 2d-2010 WL 1269696 (E.D.N.Y., April 5, 2010); *In re Isaac J.*, 705, A.D. 3d 506-N.Y.S. 2d-2010 WL 2674471 (July 6, 2010).

42. *McCarthy v. Ozark School District*, 359 F. 3d 1029 (C.A. 8 (Ark.), 2004).

43. For examples, see articles and news pieces such as Sataline, Suzanne, "A Child's Death and a Crisis of Faith," *Wall Street Journal*, June 12, 2008, http://online. wsj.com/article/SB121322824482066211.html (accessed on May 18, 2010); "Homicide Charges for Parents who Prayed as Daughter Died" ABC News, April 28, 2008, http://abcnews.go.com/Health/Story?id=4741392&page=1 (accessed on August 15, 2010); Glauber, Bill, "Parents Prayed with Ministry Founder," *Journal Sentinel* Milwaukee, WI (March 28, 2008), http://www.jsonline.com/news/wisconsin/29556439.html (accessed on August 15, 2010).

44. *Jensen v. Wagner*, 603 F.3d 1182, (C.A.10 (Utah), 2010).

45. Centers for Disease Control and Prevention, "Possible Side-effects from Vaccines," http://www.cdc.gov/vaccines/vac-gen/side-effects.htm (accessed August 15, 2010).

46. John, T. Jacob and Rueben Samuel, "Herd Immunity and Herd Effect: New Insights and Definitions," *European Journal of Epidemiology* 16 no. 7 (2002): 601-606.

47. Jacob and Samuel, "Herd Immunity."

48. See section on Compulsory Immunization in this chapter.

49. *Workman v. Mingo County Schools*, 667 F.Supp. 2d 679 (S.D.W.V. 2009).

50. For example, see American Medical Association, "AMA Signs onto Open Statement on Vaccines," http://www.ama-assn.org/ama/pub/physician-resources/public-health/vaccination-resources/pediatric-vaccination/ama-signs-open.shtml (accessed August 16, 2010).

51. American Academy of Pediatrics, "Children's Health Topics: Immunizations/Vaccines," http://www.aap.org/healthtopics/immunizations.cfm (accessed August 16, 2010).

52. Strathan, Kathleen, Alicia Gable, and Marie C. McCormick, eds., *Immunization Safety Review: Thimerosal-Containing Vaccines and Neurodevelopmental Disorders* (Washington, D.C: National Academy Press, 2001).

53. See Aster, Judith, *The Right to Health: A Resource Manual for NGO's*. Co-published by the Commonwealth Medical Trust, the American Academy for the Ad-

vancement of Science, and Human Rights Information and Documentation Systems, International, with support from the International Federation of Health and Human Organizations (Washington, D.C.: 2004, p. 21).

54. Omer, et al., "Nonmedical exemptions," 2006.

55. Salmon, Daniel A., Michael Haber, Eugene J. Gangarosa, Lynelle Phillips, Natalie J. Smith, and Robert T. Chen, "Health Consequences of Religious and Philosophical Exemptions From Immunization Laws: Individual and Societal Risk of Measles." *The Journal of the American Medical Association* 282 no. 1 (July 7, 1999): 47-53.

56. Felkin, Daniel R., Dennis C. Lezotte, Richard F. Hamman, Daniel A. Salmon, Robert T. Chen, and Richard E. Hoffman, "Individual and Community Risks of Measles and Pertussis Associated With Personal Exemptions to Immunization." *The Journal of the American Medical Association* 284 no. 24 (December 27, 2000): 3145-3150.

# Bibliography

Abaid, L.N., D.A Grimes, and K.F Schulz. "Reducing publication bias through trial registration." *Obstet Gynecol* 109 (2007): 1434-7.

Abel D. "Severin suspended for comments about Mexican immigrants." *The Boston Globe*, May 1, 2009.

Abt, Robert, Rex Schaberg, and George Hess. "Working Paper No. 4. Forest Resource Trends and Projections for North Carolina." In *Economic and ecologic impacts associated with wood chip production in North Carolina, Volume II,* Southern Center for Sustainable Forests, 2000. Http://www.env.duke.edu/scsf (accessed August 18, 2010).

Addor, M. L., T. D. Cobb, E. F. Dukes, M. Ellerbrock, and S. L. Smutko. "Linking Theory to Practice: A Theory of Change Model of the Natural Resources Leadership Institute." *Conflict Resolution Quarterly* 23, no. 2 (2005): 203-23.

Agrawal, Arun and Kent Redford. "Conservation and Displacement: An Overview." *Conservation and Society* 7 no. 1 (2009): 1-10.

Al-Jazari, Ibn al-Razzaz. *The Book of Knowledge of Ingenious Mechanical Devices, (Kitab fi Ma`rifat al-Hiyal al-Handasiyya),* translated and annotated by D. R. Hill. Dordretch, Holland: R. Reidel Publishing Company, 1974.

Aimers, James J., and Prudence M. Rice. "Astronomy, Ritual, and the Interpretation of Maya 'E-group' Architectural Assemblages." *Ancient Mesoamerica* 17 (2006): 79-96.

Allen, G. M., and E. M. Gould. "Complexity, Wickedness, and Public Forests." *Journal of Forestry* 84, no. 4 (1986): 20-23.

Altman D. G., I. Simera, J. Hoey, D. Moher, and K. Schulz. "EQUATOR: reporting guidelines for health research." *Lancet* 371 (2008): 1149–1150.

AMSA. *PharmFree Scorecard 2008.* Reston, VA: AMSA, 2008. http://amsascorecard.org/ (accessed July 10, 2010).

Anderson, Elizabeth. "Uses of value judgments in science." *Hypatia* 19, no. 1 (2004): 1-24.

Anstey, Peter. "Boyle on Occasionalism: An Unexamined Source." *Journal of the History of Ideas* 60, no. 1 (Jan 1999): 57-81.

Aquinas, Thomas. *The Works of St. Thomas Aquinas.* Second Edition. Translated by the Fathers of the English Dominican Province. London: Burns Oates and Washbourne, 1920.

Ariew, R. *Descartes and the Last Scholastics.* Ithaca: Cornell University Press, 1999.

Aruna, P. B. and Frederick W. Cubbage. "Working Paper No. 9. Regional Economic Analyses of the Forest Products and Tourism Sectors in North Carolina." In *Economic and ecologic impacts associated with wood chip production in North Carolina, Volume II,* Southern Center for Sustainable Forests, 2000. Http://www.env.duke.edu/scsf (accessed August 18, 2010).

Ashmore, Wendy and Arthur Knapp, eds. *Archaeologies of Landscape: Contemporary Perspectives.* Malden: Wiley-Blackwell, 1999.

Ashworth, William B. "Natural history and the emblematic world view." Pp. 303-322 in *Reappraisals of the Scientific Revolution*, Edited by David C. Lindberg and Robert S. Westman. Cambridge: Cambridge University Press, 1990.

Association of American Medical Colleges (AAMC). *Task Force on Financial Conflicts of Interest in Clinical Research. Protecting Subjects, Preserving Trust, Promoting Progress. Guidelines for Dealing with Faculty Conflicts of Commitment and Conflicts of Interest in Research.* Washington, DC. AAMC, 2001.

Association of American Universities (AAU). *Task Force on Research Accountability. Report on Individual and Institutional Financial Conflict of Interest.* Washington, D.C.: AAU, 2001.

Atkinson, M. "'Peer Review' Culture." *Science and Engineering Ethics* 7 (2001): 193-204.

Augustine of Hippo. *The Works of Augustine*. Marcus Dods, editor and translator. Edinburgh: T. & T. Clark, 1871.

Bacon, Francis. *The Works of Sir Francis Bacon*. London: Bayne & Son, 1844.

Bacon, Francis. *Novum Organum,* Book I, Aphorisms 1-68, 1620.

Ballard, R. (ed.) *Archaeological Oceanography*. Princeton: Princeton University Press, 2008.

Barker, Kristin. *The Fibromyalgia Story: Medical Authority and Women's Worlds of Pain*. Philadelphia: Temple University Press, 2005.

Barry, Dwight and Max Oelschlaeger. "A Science for Survival: Values and Conservation Biology." *Conservation Biology* 10 no. 3 (1996): 905-911.

Barry, J. *The Great Influenza: The Epic Story of the Deadliest Plague in History*. New York: Viking, 2004.

Bass, G.F. *Archaeology Under Water*. New York: Praeger, 1966.

———., ed. *Beneath the Seven Seas*. London: Thames & Hudson, 2005.

Bass, G.F. and F.H. van Doorninck. *Yassi Ada I. A Seventh Century Byzantine Shipwreck*. College Station: Texas A&M University Press, 1982.

Bass, G.F., Lledo, B., Matthews, S. and Brill, R.H. *Serçe Limani, Vol 2: The Glass of an Eleventh-Century Shipwreck* (Ed Rachal Foundation Nautical Archaeology Series). College Station: Texas A&M University Press, 2009.

Beecher, N., E. Harrison, N. Goldstein, M. McDaniel, P. Field, and L. Susskind. "Risk Perception, Risk Communication, and Stakeholder Involvement for Biosolids Management and Research." *Journal of Environmental Quality* 34, no. 1 (2005): 122-28.

Bell, James. *Reconstructing prehistory: Scientific method in archaeology*. Philadelphia: Temple University Press, 1994.

Bentley, Richard. *The Works of Richard Bentley*. Edited by A. Dyce. Francis McPherson: London, 1836.

Bhandari, M., J.W. Buss, D. Jackowski, V.M. Montori, H. Schünemann, S. Sprague, D. Mears, E.H. Schemitsch, D. Heels-Ansdell, and P.J. Devereaux. "Association between industry funding and statistically significant pro-industry findings in medical and surgical randomized trials." *CMAJ* 170, no. 4 (Feb 17, 2004): 477-80

Binford, Lewis. "Archaeology as Anthropology." *American Antiquity* 27, no. 4 (1962): 217-225.

———. "Smudge Pits and Hide Smoking: The Use of Analogy in Archaeological Reasoning." *American Antiquity* 32, no. 1 (1967): 1-12.

———. "Archaeological Reasoning and Smudge Pits-Revisited." Pp. 52-58 in *An Archaeological Perspective*, edited by Lewis Binford. New York: Seminar Press, 1972.

Bird, Chloe E., and Patricia P. Rieker. *Gender and health: The effects of constrained choices and social policies*. Cambridge: Cambridge University Press, 2008.

Blackstock, K. L., and C. E. Carter. "Operationalising Sustainability Science for a Sustainability Directive? Reflecting on Three Pilot Projects." *Geographical Journal* 173 (2007): 343-57.

Boettiger, S., A.B. Bennett, and Bayh-Dole. "If We Knew Then What We Know Now." *Nat Biotechnol*. 24, no. 3 (March 2006): 320-3.

Bombardier et. al., "Comparison of upper gastrointestinal toxicity of rofecoxib and naproxen in patients with rheumatoid arthritis." *New England Journal of Medicine* 353, no. 26 (2005): 2813-4.

Boyd, E.A and L.A. Bero. "Defining financial conflicts and managing research relationships: an analysis of university conflict of interest committee decisions." *Science and Engineering Ethics* 13, no. 4 (2007): 415–435.

Boyd, E.A., M.K. Cho, and L.A. Bero. "Financial conflict-of-interest policies in clinical research: issues for clinical investigators." *Acad Med* 78 (2003): 769-74.

Boyle, Robert. *The Works of Robert Boyle*. Edited by Michael Hunter and Edward Davis. London: Pickering & Chatto, 2000.

British Broadcasting Corporation (BBC) News. "Angola's town of fear and lonely death." May 3, 2005.

Brody, B. "Intellectual property and biotechnology: the U.S. internal experience-Part I." *Kennedy Inst Ethics J*. 16, no. 1(2006): 1-37.

Brunner, R., and T. A. Steelman. "Beyond scientific management." Pp. 1-46 in *Adaptive governance: integrating science, policy, and decision making*, by R. Brunner, T. A. Steelman, L. Coe-Juell, C. M. Cromley, C. M. Edwards, and D. W. Tucker. New York, Columbia University Press. 2005.

Budgell, E. *Memoirs of the lives and characters of the illustrious family of the Boyles*. Third Edition. London: Oliver Payne, 1737.

Bunimovitz, Shlomo and Raphael Greenberg (2006). "Of Pots and Paradigms: Interpreting the Intermediate Bronze Age in Israel/Palestine." Pp. 23-32 in *Confronting the Past: Archaeological and Historical Essays on Ancient Israel in Honor of William G. Dever*, edited by Seymour Gitin. Winona Lake: Eisenbrauns, 2006.

Cain, Daylian M. and Allan S. Detsky. "Everyone's a Little Bit Biased (Even Physicians)." *JAMA*. 299, no. 24 (2008): 2893-2895.

Cain, Daylian M., G. Loewenstein, and D.A. Moore. "The dirt on coming clean: perverse effects of disclosing conflicts of interest." *J Legal Stud*. 34 (2005): 1-25.

Calaprice, A., ed. *The Expanded Quotable Einstein*. Princeton, NJ: Princeton University Press, 2000.

Caldera, Y. M., A. C. Huston, and M. O'Brien. "Social interactions and play patterns of parents and toddlers with feminine, masculine, and neutral toys." *Child Development* 60 (1989): 70–76.

Campbell, E.G., B.R. Clarridge, M. Gokhale, et. al. "Data withholding in academic genetics: evidence from a national survey." *JAMA* 287 (2002): 473-80.

Campbell, E.G. "Doctors and drug companies—scrutinizing influential relationships." *New England Journal of Medicine* 357, no. 18 (2007): 1796-7.

Campbell, Philip. "Introducing a new policy for authors of research papers in *Nature* and *Nature* journals." *Nature* 412, 751 (August 23, 2001): 751.

———. Editorial Note. *Nature* 416, no. 6881 (April 11, 2002): 600.

———. "Declaration of financial interests." *Nature* 412 (2001):751.

Canfora, L. *The Vanished Library: A Wonder of the Ancient World* (Hellenistic Culture and Society) Berkeley: University of California Press, 1990.

Caplan, A. "Halfway there: the struggle to manage conflicts of interest." *J. Clin. Invest*. 117 (2007): 509-510.

Carlson, D.N. "Caligula's Floating Palaces." *Archaeology* 55, no. 3 (May/June 2002): 26.

Carrier, Martin, Don Howard, and Janet Kourany, eds. *The Challenge of the Social and the Pressure of Practice*. Pittsburgh, University of Pittsburgh Press, 2008.

Cartwright, Nancy. *How the Laws of Physics Lie*. New York: Oxford University Press, (1983).

Cash, D. W., W. C. Clark, F. Alcock, N. M. Dickson, N. Eckley, D. H. Guston, J. Jager, and R. B. Mitchell. "Knowledge Systems for Sustainable Development." *Proceedings of the National Academy of Sciences of the United States of America* 100, no. 14 (2003): 8086-91.

Casson, L. *Ships and Seamanship in the Ancient World.* Princeton: Princeton University Press, 1971.

Centers for Disease Control and Prevention. "Outbreak of Ebola viral hemorrhagic fever—Zaire, 1995." *Morbidity and Mortality Weekly Report* 44, no. 19 (May 19, 1995): 381-2.

———. "Outbreak of Marburg virus hemorrhagic fever—Angola, October 1, 2004-March 29, 2005." *Morbidity and Mortality Weekly Report*, 54 no. 12 (Apr 1, 2005): 308-9.

———. "Outbreak of Acute Illness—Southwestern United States, 1993." *Morbidity and Mortality Weekly Report* 42, no. 22 (June 11, 1993):421-4.

Chang, Hui-hua. *Testing the Serpent of Asclepius: The Social Mobility of Greek Physicians* (PhD dissertation, Indiana University, 2003), 152.

Charlton, B.G. "Conflicts of interest in medical science: peer usage, peer review and 'Col consultancy'." *Medical Hypotheses* 63, no. 2 (2004): 181-186.

Charatan, F. "Doctors say they are not influenced by drug companies' promotions." *BMJ* 322 (2001): 1081.

Christou, Paul. "No credible scientific evidence is presented to support claims that transgenic DNA was introgressed into traditional maize landraces in Oaxaca, Mexico." *Transgenic Research*, 11 (2002): iii-v.

Clark, Tim W. *The Policy Process: A Practical Guide for Natural Resource Professionals*. New Haven, CT: Yale University Press, 2002.

Clarke, A. C. *Profiles of the Future: An inquiry into the limits of the possible*. New York: Harper & Row, 1973.

Clarke, David *Analytical Archaeology*. London, Methuen Press, 1968.

Clough, Sharyn. *Beyond Epistemology: A Pragmatist Approach to Feminist Science Studies*. Lanham, MD: Rowman and Littlefield, 2003.

Colhoun, H.M., P.M. McKeigue, and G.D. Smith. "Problems of reporting genetic associations with complex outcomes." *Lancet* 361 (2003): 865–72.

Collins, F. S. *The Language of God: A Scientist Presents Evidence for Belief.* New York: Free Press, 2006.

Contreras J. "On Scene in the Hot Zone." *Newsweek*. May 29, 1995.

Cook, R. and E. Calabrese. "The importance of hormesis to public health." *Environ Health Perspect* 114 (2006) :1631–1635.

Cook, Margaret. "Divine Artifice and Natural Mechanism." *Osiris* 2nd Series 16 (2001): 133-150.

Collins, H.M. and R. Evans. "The Third Wave of Science Studies: Studies of Expertise and Experience." *Social Studies of Science* 32, no. 2 (2002): 235-96.

Corburn, Jason. *Street Science: Community Knowledge and Environmental Health Justice*, Urban and Industrial Environments. Cambridge, MA: MIT Press, 2005.

Cottingham, Stoothoff and Murdoch. trans. *Philosophical Writings of Descartes* (3 vol), Cambridge: Cambridge University Press, 1995.

Crasnow, Sharon. "Activist research and the objectivity of science." *American Philosophical Newsletter on Feminism and Philosophy* 6 no. 1 (2006): 3-5.

Crick, F. *What Mad Pursuit: A Personal View of Scientific Discovery*. New York: Harper Collins, 1988.

Cubbage, F. W. "Costs of Forestry Best Management Practices: A Review." *Water, Air, and Soil Pollution: Focus* 4, no. 1 (2004): 131-42.

Cubbage, F. W., W. Scott Burleson, and John D. Dodrill. "Working Paper No. 1: North Carolina's Forests. Trends from 1938 to 1990." In *Economic and ecologic impacts associated with wood chip production in North Carolina, Volume II,* Southern Center for Sustainable Forests, 2000. Http://www.env.duke.edu/scsf (accessed August 18, 2010).

Culliton, B.J. "Coping with fraud: The Darsee Case." *Science,* 220 (1983): 31-35.

Curfman, G.D., S. Morrissey, and J.M. Drazen. "Expression of concern." *New England Journal of Medicine* 343 (2000): 1520-8.

Dalton, R. "Postdocs slam zealous attitude of NIH ethics office." *Nature* 434 (2005): 687.

Dawkins, R. *The God Delusion.* New York: Bantam Books, 2006.

Davis, D. *Commercial Navigation in the Greek and Roman World* (PhD. Dissertation, The University of Texas at Austin), 2009.

Davis, Matthew. "Parents May Underestimate the Risks of H1N1 Flu for Their Children," *CS Mott Children's Hospital National Poll on Children's Health* 8, no. 1 (September 24, 2009).

de Melo-Martín, I. and K. Intemann. "How do disclosure policies fail? Let us count the ways." *FASEB J.* 23, no. 6 (2009): 1638-42.

DeAngelis, C.D., P.B. Fontanarosa, and A. Flanagin. "Reporting financial conflicts of interest and relationships between investigators and research sponsors." *JAMA* 286 (2001): 89-91.

Dean-Jones, Lesley. "Literacy and the Charlatan n Ancient Greek Medicine." In *Written Texts and the Rise of Literate Culture in Ancient Greece,* ed. Harvey Yunis. Cambridge: Cambridge University Press, 2003.

Delgado, J. (ed.) *Encyclopedia of Underwater and Maritime Archaeology.* New Haven: Yale University Press, 1998.

Dennett, D. C. *Breaking the Spell: Religion as a Natural Phenomenon.* New York: Viking (Penguin), 2006.

———. "Thank Goodness!" *Edge: The Third Culture.* Edge Foundation, Inc., 2006. http://edge.org/third_culture/dennett06/dennett06_index.html (accessed August 19, 2010).

Descartes, René. *The Philosophical Writings of Descartes.* Vol. 1. Translated and edited by John Cottingham, Robert Stoothoff and Dugald Murdoch. Cambridge: Cambridge University Press, 1985.

Desmond, A. and J. Moore. *Darwin: The Life of a Tormented Evolutionist.* New York: W. W. Norton & Company, 1992.

Dever, William. "The Impact of 'New Archaeology' on Syro-Palestinian Archaeology." *BASOR,* number 242 (Spring 1981): 15-30.

———. "Syro-Palestinian and Biblical Archaeology." Pp. 31-74 in *The Hebrew Bible and its Modern Interpreters.* Edited by Douglas Knight and Gene Tucker. Philadelphia: Fortress Press, 1985.

———. *What Did the Biblical Writers Know & When Did They Know It?* Grand Rapids: William B. Eerdmans Publishing Company, 2001.

Dodrill, John D., and Frederick W. Cubbage. "Working Paper No. 3. Potential Wood Chip Mill Harvest Area Impacts in North Carolina." In *Economic and ecologic impacts associated with wood chip production in North Carolina, Volume II,* Southern Center for Sustainable Forests, 2000. Http://www.env.duke.edu/scsf (accessed Auust 18, 2010).

Dodrill, J. D., F. W. Cubbage, R. H. Schaberg, and R. C. Abt. "Wood Chip Mill Harvest Volume and Area Impacts in North Carolina." *Forest Products Journal* (2002): 29-37.

Dorsey E.R., J. de Roulet, J.P. Thompson, J.I. Reminick, A. Thai, Z. White-Stellato, C.A. Beck, B.P. George, and H. Moses third. "Funding of US biomedical research, 2003-2008." *JAMA* 3032 (2010): 137-43.

Douglas, H. "Inductive risk and values in science." *Philosophy of Science* 67, no. 4 (2000): 559-579.

———. The Irreducible Complexity of Objectivity. *Synthese* 138 no. 3 (2004): 453-473.

Drazen J.M., and G.D Curfman. "Financial associations of authors." *New England Journal of Medicine* 346 (2002):1901-2.

Edltorial. "Dealing with disclosure." *Nat Med* 12 (2006): 979.

Ehringhaus S. and D. Korn. "U.S. Medical School Policies on Individual Financial Conflicts of Interest." Results of an AAMC Survey. Washington, DC: AAMC, 2004.

Elliott D.E., R.W. Summers, and J.V. Weinstock. "Helminths as governors of immune-mediated inflammation." *Int Journal of Parasitology* 37, no. 5 (2007.): 457-64.

Elliott K.C. "Scientific judgment and the limits of conflict-of-interest policies." *Account Res.* 15 no. 1 (2008): 1-29.

———. "Hormesis, ethics, and public policy: an overview." *Hum Exp Toxicol.* 27 (2008): 659–662.

Epstein, S. *Inclusion: The politics of difference in medical research*. Chicago: University of Chicago Press, 2007.

Erb, K. "Can helminths or helminth-derived products be used in humans to prevent or treat allergic diseases?" *Trends in Immunology* 30, no. 2 (2009): 75-82.

Fausto-Sterling, Anne. *Myths of Gender: Biological Theories About Women and Men*, 2nd ed. New York: Basic Books, 1992.

———. "Gender identification and Assignment in Intersex Children." *Dialogues in Pediatric Urology* 25, no. 6 (2002): 4-5.

Felkin, Daniel R., Dennis C. Lezotte, Richard F. Hamman, Daniel A. Salmon, Robert T. Chen, and Richard E. Hoffman. "Individual and Community Risks of Measles and Pertussis Associated With Personal Exemptions to Immunization." *The Journal of the American Medical Association* 284, no. 24 (December 27, 2000): 3145-3150.

Fénelon, François. *Oeuvres Philosophiques de Fénelon*. Nouvelle Edition. Edited by M. A. Jacques. Paris: Charpentier, 1845.

Feyerabend, Paul. *Against Method: Outline of an Anarchistic Theory of Knowledge*. London: Humanities Press, 1975.

Forster, Malcolm and Elliott Sober. "How to tell when simpler, more unified, or less ad hoc theories will provide more accurate predictions." *British Journal for the Philosophy of Science* 45, no. 1 (1994): 1-35.

Foster, Kenneth R. and Peter Huber. 1999. *Judging Science: Scientific Knowledge and the Federal Courts*. Cambridge: MIT Press.

Freeth, T., A. Jones, J.M. Steele, and Y. Bitsakis. "Calendars with Olympiad display and eclipse prediction on the Antikythera Mechanism," *Nature* 454, (July 31 2008): 614-617.

Funtowicz, S. O., and J. R. Ravetz. "A New Scientific Methodology for Global Environmental Issues." In *Ecological Economics*, edited by R. Costanza, 137-52. New York: Columbia University Press, 1991.

Gao F., E. Bailes, and D. Robertson. "Origin of HIV" *Nature* 397, no. 6718 (Feb 4 1999): 436-41.

Garber, D. "Science and Certainty in Descartes," Pp. 114-150 in *Descartes: Critical and Interpretive Essays*, edited by Michael Hooker. Baltimore: Johns Hopkins University Press, 1978.

———. *Descartes' Metaphysical Physics*. Chicago: University of Chicago Press, 1992.

Garrett L, *The Coming Plague: Newly Emerging Diseases in a World Out of Balance*. New York: Farrar, Straus and Giroux, 1994.

Gaukroger, S., J. Schuster, and J. Sutton, eds. *Descartes' Natural Philosophy*, London: Routledge, 2000.

Gaukroger, S. *Descartes' System of Natural Philosophy*, Cambridge: Cambridge University Press, 2002.

Geiger, R., & C. Sa. *Tapping the riches of science. Universities and the promise of economic growth.* Cambridge, MA: Harvard University Press, 2008.

Gieryn, Thomas F. *Cultural Boundaries of Science: Credibility on the Line.* Chicago: University of Chicago Press, 1999.

Gitin, Seymour and William Dever, eds. *AASOR 49: Recent Excavations in Israel: Studies in Iron Age Archaeology.* Winona Lake: Eisenbrauns, 1989.

Glickman, L.T., A.O Camara, N.W Glickman, and G.P. McCabe. "Nematode Intestinal Parasites of Children in Rural Guinea, Africa: Prevalence and Relationship to Geophagia." *International Journal of Epidemiology* 28 (1999): 169-174.

Gould, S. J. "Nonoverlapping Magisteria." *Natural History* 106 (1997): 16-22.

Grant, E. *God and Reason in the Middle Ages.* Cambridge: Cambridge University Press, 2001.

Grant, E. and J. E. Murdoch, eds. *Mathematics and its Applications to Science and Natural Philosophy in the Middle Ages.* Cambridge: Cambridge University Press, 1987.

Grant, M. *Cleopatra—A Biography.* Book Sales: Northfield, 2004 (first published 1972).

Greco, Brandon, F., and Gregory, James D. "Working Paper No. 7. Storm Water and Process Water Management at North Carolina Wood Chip Mills." In *Economic and ecologic impacts associated with wood chip production in North Carolina, Volume II,* Southern Center for Sustainable Forests, 2000. Http://www.env.duke.edu/scsf (accessed August 18, 2010).

Green, P. *Alexander to Actium: The Historical Evolution of the Hellenistic Age* (Hellenistic Culture and Society). Berkeley: University of California Press, 1990.

Gross, Alan. Personal email, Sun, Jun 22 2003 06:42:58-0400 (EDT).

———. "Peer Review and Scientific Knowledge." Pp. 129-143 in *The Rhetoric of Science*, Cambridge, MA: Harvard University Press, 1990.

Guerlac, Henry and M. C. Jacob. "Bentley, Newton, and Providence: The Boyle Lectures Once More." *Journal of the History of Ideas* 30, no. 3 (Jul.-Sep., 1969): 307-318.

Habu, Junko, Clare Fawcett, and John Matsunaga, eds. *Evaluating Multiple Narratives.* New York: Springer, 2008.

Hale, John R. *Lords of the Sea: The Epic Story of the Athenian Navy and the Birth of Democracy.* New York: Penguin, 2010.

Hankin, Benjamin L., and Lyn Abramson. "Development of gender differences in depression: An elaborated cognitive vulnerability—transactional stress theory." *Psychological Bulletin* 127, no. 6 (2001): 773-796.

Hanson, N. R. "The Agnostic's Dilemma." Pp. 303-308 in *What I do not believe and other essays*, edited by Toulmin, S. and H. Woolf. Dordrecht: D. Reidel Publishing Company, 1971.

———. "What I Don't Believe." Pp. 309-331 in *What I do not believe and other essays*, edited by Toulmin, S. and H. Woolf. Dordrecht: D. Reidel Publishing Company, 1971.

Hanson, N. R., S. Toulmin, et al. *What I do not believe, and other essays.* Dordrecht: D. Reidel Publishing Company, 1971.

Harding, Sandra. "Rethinking Standpoint Epistemology: What is 'Strong Objectivity'?" Pp. 49-82 in *Feminist Epistemologies*, edited by Linda Alcoff and Elizabeth Potter. NY and London: Routledge, 1993.

———. *The Science Question in Feminism.* Ithaca: Cornell University Press, 1986.

———. *Whose Science? Whose Knowledge?: Thinking from Women's Lives.* Ithaca, N.Y.: Cornell University Press, 1991

————. (ed). *The Racial Economy of Science: Toward a Democratic Future.* Bloomington, IN: Indiana University Press, 1993.

Harris, S. *The End of Faith: Religion, Terror, and the Future of Reason.* New York:
W.W. Norton, 2004.

————. *Letter to a Christian Nation.* New York: Random House, 2006.

Harrison, C. "Peer review, politics and pluralism." *Environmental Science & Policy* 7,
no. 5 (2004): 357-368.

Hatfield, G. "Science, Certainty and Descartes." Pp. 249-262. in A. Fine and J. Leplin,
eds. *Philosophy of Science Association* vol. 2. East Lansing, MI: Philosophy of Science Association. 1988.

————. *Routledge Guide to Descartes' Meditations.* London: Routledge, 2002.

Hayward, Steven F. "Environmental Science and Public Policy." *Social Research* 73, no.
3 (2006): 891-914.

Heller, John L. "Notes on the Titulature of Linnaean Dissertations." *Taxon* 32, no. 2,
(May, 1983): 218-252.

Hemingway, S. *The Horse and Jockey from Artemision: A Bronze Equestrian Monument
of the Hellenistic Period* (Hellenistic Culture and Society). Berkeley: University of
California Press, 2004.

Hempel, Carl. "The Function of General Laws in History." *Journal of Philosophy* 39,
no.2 (1942): 35-48.

————. "Science and Human Values." in *Aspects of Scientific Explanation*, New York:
Free Press, 1965.

————. *The Philosophy of Natural Science.* Upper Saddle River: Prentice Hall, 1966.

Hempel, Carl and Paul Oppenheim. "Studies in the Logic of Explanation." *Philosophy of
Science* 15, no. 2 (1948): 135-175.

Hess, George, Stacy Sherling, Robert Abt, and Rex Schaberg. "Working Paper No. 6,
Part I: Trends in Forest Composition and Size Class Distribution: Implications for
wildlife habitat." In *Economic and ecologic impacts associated with wood chip production in North Carolina, Volume II,* Southern Center for Sustainable Forests,
2000. Http://www.env.duke.edu/scsf (accessed August 18, 2010).

Hess, George and Dale Zimmerman. "Woody Debris Volume on Clearcuts with and
without Satellite Chip Mills." *Southern Journal of Applied Forestry* 25, no. 4
(2000): 173-77.

————. "Working Paper No. 6, Part II: The effect of satellite chip mills on post-harvest
woody debris." In *Economic and ecologic impacts associated with wood chip production in North Carolina, Volume II,* Southern Center for Sustainable Forests,
2000. Http://www.env.duke.edu/scsf (accessed August 18, 2010).

Hill, D.R. *A History of Engineering in Classical and Medieval Times.* New York:
Routledge, 1996.

Hirsch L.J. "Conflicts of interest, authorship, and disclosures in industry-related scientific
publications: the tort bar and editorial oversight in medical journals." *Mayo Clinic
Proceedings* 84, no. 9 (2009): 811-821.

Hitchcock, C. *Contemporary Debates in Philosophy of Science.* Oxford: Blackwell Publishers, 2004.

Hodder, Ian. *The Archaeological Process: An Introduction.* Oxford: Blackwell Publishers, 1999.

————. *Toward a Reflexive Method in Archaeology: The Example at Çatalhöyük.* Cambridge: British Institute of Archaeology at Ankara, 2000.

————. "Multivocality and Social Archaeology." Pp. 196-200 in *Evaluating Multiple
Narratives.* New York: Springer, 2008.

Hooke, Robert. *Micrographia.* London: Royal Society, 1665.

Horton R. "Conflicts of interest in clinical research: opprobrium or obsession?" *Lancet* 349 (1997): 1112-3.

———. "Expression of concern: non-steroidal anti-inflammatory drugs and the risk of oral cancer." *Lancet* 367, no. 9506 (2006): 196.

Howes, Moira. 2007. "Maternal Agency and the Immunological Paradox of Pregnancy." Pp. 179-198 in *Establishing Medical Reality: Essays in the Metaphysics and Epistemology of Biomedical Science*, edited by Harold Kincaid and Jennifer McKitrick. Heidelberg: Springer Netherlands, 2007.

Hull, David L. *Science as a Process*. Chicago: University of Chicago Press, 1988.

———. "Why Scientists Behave Scientifically." Pp. 135-138 in *Science and Selection: Essays on Biological Evolution and the Philosophy of Science*. New York: Cambridge University Press, 2001.

Humphrey, J.W., J.P. Oleson, and A.N. Sherwood. *Greek and Roman Technology: A Sourcebook: Annotated Translations of Greek and Latin Texts and Documents* (Routledge Sourcebooks for the Ancient World). New York: Routledge, 1998.

Hurtado, A., K. Hill, I. Arenas de Hurtado, and Selva Rodriguez. "The evolutionary context of chronic allergic conditions: The Hiwi of Venezuela." *Human Nature* 8, no.1 (1997): 51-75.

Huxley, T. *Collected Essays*. Thoemmes Continuum; Facsimile of 1893-4 edition, 2001.

Institute of Medicine (IOM). *Conflict of interest in medical research, education, and practice*. Washington, D.C.: National Academies Press, 2009.

Intemann, Kristen. "Science and Values; Are Moral Judgments Always Irrelevant to the Justification of Scientific Claims?" *Philosophy of Science* 68, no. 3 (2001): S506-18.

Ioannidis, J.P., E.E. Ntzani, T.A. Trikalinos, and D.G. Contopoulos-Ioannidis. "Replication validity of genetic association studies." *Nat Genet*. 29, NO. 3 (Nov 2001): 306-9.

Israel, Jonathan. *Radical Enlightenment*. Oxford: Oxford University Press, 2001.

Jacobson, D.L., S.J. Gange, N.R. Rose, and N.M. Graham. "Epidemiology and estimated population burden of selected autoimmune diseases in the United States." *Clin Immunol Immunopathol* 84, no. 3 (1997): 223-243.

Jahrling P.B., T.W. Geisbert, D.W. Dalgard et al. "Isolation of Ebola virus from imported monkeys in the United States." *Lancet* 335 (1990): 502-5.

James, P. and N. Thorpe. *Ancient Inventions*. New York: Ballantine Books, 1995.

John, T. Jacob and Rueben Samuel. "Herd Immunity and Herd Effect: New Insights and Definitions." *European Journal of Epidemiology* 16 no. 7 (2002): 601-606.

Johnson, C., E. Peterson, and Dennis R. Ownby. "Gender Differences in Total and Allergen-specific Immunoglobulin E (IgE) Concentrations in a Population-based Cohort from Birth to Age Four Years." *Am. J. Epidemiology* 147(1998): 1145-1152.

Kaiser, J. "Conflict of interest. NIH rules rile scientists, survey finds." *Science* 314 (2006): 740.

Kaiser, J. "Private money, public disclosure." *Science* 325 (2009): 28–30.

Kaplinsky, Nick, et al., "Biodiversity: Maize transgene results in Mexico are artifacts." *Nature* 416, no. 6881 (April 11 2002): 601.

———. "Conflicts around a study of Mexican crops; Kaplinsky replies." *Nature* 416 (June 27, 2002): 898.

Karl, H. A., L. E. Susskind, and K. H. Wallace. "A Dialogue Not a Diatribe—Effective Integration of Science and Policy through Joint Fact Finding." *Environment* (2007): 20.

Kates, R. W., W. C. Clark, R. Corell, J. M. Hall, C. C. Jaeger, I. Lowe, J. J. McCarthy, H. J. Schellnhuber, B. Bolin, N. M. Dickson, S. Faucheux, G. C. Gallopin, A. Grubler, B. Huntley, J. Jager, N. S. Jodha, R. E. Kasperson, A. Mabogunje, P. Matson, H. Mooney, B. Moore, T. O'Riordan, and U. Svedin. "Environment and Development-

Sustainability Science." *Science* (2001): 641-42.

Katz D., A.L. Caplan, and J.F. Merz. "All gifts large and small: toward an understanding of the ethics of pharmaceutical industry gift-giving." *American Journal of Bioethics* 3, no. 3 (2003): 39-46.

Keeley, B. L., ed. *Paul Churchland. Contemporary Philosophy in Focus.* New York: Cambridge University Press, 2005.

———. "God as the ultimate conspiracy theory." *Episteme: A Journal of Social Epistemology* 4, no. 2 (2007): 135-149.

Keynes, M. "The Portland Vase: Sir William Hamilton, Josiah Wedgwood and the Darwins," *Notes and Records of the Royal Society of London* 52, no. 2 (Jul., 1998): 237-259.

Khaiboullina S., S. Morzuno, and S. St Jeor. "Hantaviruses: molecular biology, evolution, and pathogenesis." *Curr Mol Med* 5, no. 8 (Dec, 2005): 773-90.

Khan S.N., M.J. Mermer, E. Myers, and H.S. Sandhu. "The roles of funding source, clinical trial outcome, and quality of reporting in orthopedic surgery literature." *American Journal of Orthopedics (Belle Mead NJ)* 37, no. 12 (Dec 2008): E205-12.

Kilpeläinen, M., E.O. Terho, H, Helenius, and M. Koskenvuo. "Farm environment in childhood prevents the development of allergies." *Clinical and Experimental Allergy* 30, no. 2 (2000): 201-208.

Kincaid, Harold, John Dupré, and Alison Wylie, eds. *Value-Free Science?: Ideals and Illusions.* New York: Oxford University Press, 2007.

Kinchy, A., and D. L. Kleinman. "Organizing Credibility: Discursive and Organizational Orthodoxy on the Borders of Ecology and Politics." *Social Studies of Science* (2003): 869-96.

Kitcher, Philip. Responsible Biology. *BioScience* 54 no. 4 (2004): 331-336.

———. *Living with Darwin: Evolution, Design, and the Future of Faith.* New York: Oxford University Press, 2007.

Kohler, Robert. *Landscapes and Labscapes: Exploring the Lab-Field Border in Biology.* Chicago and London: The University of Chicago Press, 2002.

Korfmacher, K. S. "Invisible Successes, Visible Failures: Paradoxes of Ecosystem Management in the Albemarle-Pamlico Estuarine Study." *Coastal Management* (1998): 191-211.

———. "Science and Ecosystem Management in the Albemarle-Pamlico Estuarine Study." *Ocean & Coastal Management* (2002): 277-300.

Kosso, Peter. *Knowing the Past: Philosophical Issues of History and Archaeology.* Amherst: Humanity Books, 2001.

———. "Scientific Understanding." *Foundations of Science* 12, vol. 2 (2007): 173-188.

Krieger, William. *Can There Be a Philosophy of Archaeology: Processual Archaeology and the Philosophy of Science?* Lanham: Lexington Books, 2006.

Krimsky, S. *Science in the Private Interest.* Lanham, MD: Rowman and Littlefield, 2003.

Krimsky, S. and L. Rothenberg. "Conflict of interest policies in science and medical journals: Editorial practices and author disclosure." *Science and Engineering Ethics* 7 (2001): 205–218.

Kuhn, Thomas. *The Structure of Scientific Revolutions.* Chicago: the University of Chicago Press, 1970.

———. *The essential tension: selected studies in scientific tradition and change.* Chicago: University of Chicago Press, 1977.

Kukla, A. "Forster and Sober on the Curve-Fitting Problem." *The British Journal for the Philosophy of Science* 46, no. 2, (Jun. 1995).

Lacey, Hugh. *Is Science Value-free?: Values and Scientific Understanding.* New York: Routledge, 1999.

———. Assessing the Value of Transgenic Crops. *Science and Engineering Ethics* 8 no. 4 (2002): 497-511.

Lach, D., P. List, B. Steel, and B. Shindler. "Advocacy and Credibility of Ecological Scientists in Resource Decisionmaking: A Regional Study." *Bioscience* (2003): 170-78.

Lackey, R. T. "Science, Scientists, and Policy Advocacy." *Conservation Biology* (2007): 12-17.

Lagakos, S.W. "Time-to-event analyses for long-term treatments: the APPROVe trial." *New England Journal of Medicine* 355 (2006):113-117.

Laine, C., R. Horton, C.D. DeAngelis, J.M. Drazen, F.A. Frizelle, F. Godlee, C. Haug, P.C. Hébert, S. Kotzin, A. Marusic, P. Sahni, T.V. Schroeder, H.C. Sox, M.B. Van der Weyden, and F.W. Verheugt. "Clinical trial registration: looking back and moving ahead." *JAMA*. 298, no. 1 (2007): 93-4.

Lakatos, Imre and Paul Feyerabend. *For and Against Method*. Chicago: University of Chicago Press, 1999.

Lamb, Sarah. "The politics of dirt and gender: Body techniques in Bengali India." Pp. 213-232 in *Dirt, Undress and Difference: Critical Perspectives on the Body's Surface*, edited by Adeline Masquelier. Bloomington: Indiana University Press, 2005.

Lanphear, B. and K. Roghmann. "Pathways of lead exposure in urban children." *Environmental Research* 74 (1997): 67–73.

Laskaris, *The Art is Long: On the Sacred Disease and the Scientific Tradition*. Leiden: Brill, 2002.

Lazenby, F.D. "A Note on Vitrum Flexile," *The Classical Weekly* 44, no. 7 (Jan. 15, 1951): 102-103.

Leach, Joan, *Healing and the Word: Hippocratic Medicine and Sophistical Rhetoric in Classical Antiquity*. Ph.D. dissertation, University of Pittsburgh, 1996.

Lewis, M.J.T. "The Origins of the Wheelbarrow," *Technology and Culture*, 35, no. 3. (July) 1994: 453–475.

Linnaeus, Carolus et al. *Miscellaneous Tracts relating to Natural History, Husbandry, and Physick*. Third edition. Benjamin Stillingfleet, Translator. London: J. Dodsley, 1775.

Linnaeus, Carolus et al. *Select dissertations from the Amoenitates Academicae: a supplement to Mr. Stillingfleet's tracts relating to natural history*. Translated by F. J. Brand. London: G. Robinson, 1781.

Lipton, S., E.A. Boyd, and L.A. Bero. "Conflicts of interest in academic research: policies, processes, and attitudes." *Account Res.* 11 (2004): 83-102.

Lloyd, Elisabeth A. "Objectivity and the Double Standard for Feminist Epistemologies." *Synthese* 104 no. 3 (1995):351-381.

Lloyd, G.E.R. "Hellenistic science." Pp. 321-352 in *Cambridge Ancient History*, Vol. VII Part I, second edition, 1984.

———. *In the Grip of Disease: Studies in the Greek Imagination* (New York: Oxford University Press, 2003)

Locke, John. *An Essay Concerning Human Understanding*. Edited by P. H. Nidditch. Oxford: Clarendon, 1975.

Lofton, J. "An Exchange: My Correspondence With Milton Friedman About God, Economics, Evolution And 'Values.'" *The American View* (Oct-Dec 2006).

Longino, Helen, *Science as Social: values and objectivity in scientific inquiry*. Princeton: Princeton University Press, 1990.

———. "Can there be a feminist science?" *Hypatia* 2, no. 3 (1987): 51-64.

———. *The Fate of Knowledge*. Princeton: Princeton University Press, 2002.

———. "How Values Can Be Good for Science." Pp. 127-142 in *Science, Values, and Objectivity*, edited by Machamer and Walters. Pittsburgh: University of Pittsburgh Press, 2004.

————. "Comments on Science and Social Responsibility: A Role for Philosophy of Science?" *Philosophy of Science* 64, no. 4 (1997): 179.

Loewenberg, S. "The Bayh-Dole Act: a model for promoting research translation?" *Mol Oncol.* 3, vol. 2 (2009): 91-3.

Lubchenco, J. "Entering the Century of the Environment: A New Social Contract for Science." *Science* (1998): 491-97.

Luks, F., and B. Siebenhuner. "Transdisciplinarity for Social Learning? The Contribution of the German Socio-Ecological Research Initiative to Sustainability Governance." *Ecological Economics* (2007): 418-26.

Luoma, J. R. "Whittling Dixie: Highly Automated 'Supermills' Are Devouring Vast Tracts of Hardwoods, as a New Kind of Forestry Slices through the Southland." *Audubon*, 1997, 38–45, 97–100.

Lynch, Michael. "From Ruse to Farce." *Social Studies of Science* 36 no. 6 (2006): 819-826.

Machamer, Peter and Gereon Walters, eds. *Science, Values, and Objectivity*. Pittsburgh: University of Pittsburgh Press, 2004.

Maddy, P. *Second Philosophy: A Naturalistic Method*. Oxford: Oxford University Press, 2007.

Maguire, Lynn A. "Making the Role of Values in Conservation Explicit: Values and Conservation Biology." *Conservation Biology* 10 no. 3 (1996): 914-916.

Mahlman, J. D. "Uncertainties in Projections of Human-Caused Climate Warming." *Science* (1997): 1416-17.

Maldonado, Yvonne A. "Current Controversies in Vaccination: Vaccine Safety," *Journal of the American Medical Association*, 288, no. 24 (December 25, 2002): 3155-3158.

Mallard, G., M. Lamont, and J. Guetzkow. "Fairness as appropriateness: Negotiating epistemological differences in peer review." *Science Technology Human Values*, 34, no. 5 (2009): 573-606.

Manuel, J. "Do Wood Chip Mills Threaten the Sustainability of North Carolina Forests?" *North Carolina Insight* 18 (1999): 66-93.

Marsden, E.M. *Greek and Roman Artillery: Technical Treatises*. 2 vols. Oxford: Clarendon Press, 1971.

Martin, Karin A. "Becoming a gendered body: Practices of preschools." *American Sociological Review* 63, no. 4 (1998): 494-511.

Maziak, W, T. Behrens, T. M. Brasky, H. Duhme, P. Rzehak, S. K. Weiland, and U. Keil. "Are asthma and allergies in children and adolescents increasing? Results from ISAAC phase I and phase III surveys in Munster, Germany." *Allergy* 58 (2003): 572–579.

McClary, C. D. "The 'Chip Mill' Issue: Sustainable Forestry?" (n.d.).

McCool, S. F., and K. Guthrie. "Mapping the Dimensions of Successful Public Participation in Messy Natural Resources Management Situations." *Society & Natural Resources* (2001): 309-23.

McCoy, Earl. "Advocacy as Part of Conservation Biology." *Conservation Biology* 10 no. 3 (1996): 919-920.

McDade, Thomas, Julienne Rutherford, Linda Adair, and Christopher Kuzawa. "Early Origins of Inflammation: Microbial Exposures in Infance Predice Lower Levels of C-Reactive Protein in Adulthood." *Proceedings of The Royal Society B: Biological Sciences*. http://rspb.royalsocietypublishing.org/content/early/2009/12/08/rspb.2009.1795.full (accessed July 10, 2010).

McDowell, G. R. "Engaged Universities: Lessons from the Land-Grant Universities and Extension." *Annals of the American Academy of Political and Social Science* (2003): 31-50.

McGrew, T. "Confirmation, Heuristics, and Explanatory Reasoning. " *British Journal for the Philosophy of Science* 54 (2003): 553–567.

Meffe, Gary K. "Policy Advocacy and Conservation Science: Introduction." *Conservation Biology* 21 no. 1 (2007): 11.

Meine, Curt, and Gary K. Meffe. "Conservation Values, Conservation Science: A Healthy Tension." *Conservation Biology* 10 no. 3 (1996): 916-917.

Mello, M.M., B.R. Clarridge, and D.M. Studdert. "Academic medical centers' standards for clinical-trial agreements with industry." *New England Journal of Medicine* 352, no. 21 (2005): 2202-2210.

———. "Researchers' views of the acceptability of restrictive provisions in clinical trial agreements with industry sponsors." *Account. Res.* 12, no. 3(2005): 63-91.

Metz, Matthew and Johannes Fütterer. "Biodiversity: Suspect evidence of transgenic contamination." *Nature* 416, no. 6881 (April 11, 2002): 600-601.

———. "Biodiversity: Suspect evidence of transgenic contamination; Metz replies." *Nature* 416 (June 27, 2002): 897-898.

Michaels, D. *Doubt is their product.* New York: Oxford University Press, 2008.

Miller, K. R. *Finding Darwin's God: A Scientist's Search for Common Ground Between God and Evolution.* New York: Harper Perennial, 2000.

Mithen, Steven. *The Prehistory of the Mind: A Search for the Origins of Art, Religion, and Science.* London: Thames and Hudson, Ltd, 1996.

Mornet, Daniel. *Les Sciences de la Nature en France, au XVIIIe Siècle.* Paris: Armand Colin, 1911.

Morrison, M. *Unifying Scientific Theories: Physical Concepts and Mathematical Structures.* Cambridge: Cambridge University Press, 2000.

Mowery, D., R. Nelson, B. Sampat, and A. Ziedonis. *Ivory tower and industrial innovation-University-industry technology transfer before and after the Bayh–Dole Act.* Stanford: Stanford Business Books, 2004.

Munson, Patrick. "Comments on Binford's 'Smudge Pits and Hide Smoking: The Use of Analogy in Archaeological Reasoning.'" *American Antiquity* 34, no. 1 (1969): 83-85.

Murphy FA, Kiley MP, Fisher-Hoch SP. "Filoviridae: Marburg and Ebola viruses." In *Virology*, ed. Fields BN, Knipe DM. (New York: Raven Press, Ltd., 1990), 933–42.

Murthy, Aruna, and Frederick W. Cubbage. "An Economic Analysis of Forest Products and Nature-Based Tourism Sectors in North Carolina." *Southern Rural Sociology* 20, no. 1 (2004): 25-38.

Murray, Tim. *The Encyclopedia of Archaeology: Volume 2, The Great Archaeologists.* Santa Barbara: ABC (1999).

Myrvold, W. C. "A Bayesian Account of the Virtue of Unification," *Philosophy of Science* 70 (2003): 399–423.

Nelson, Cary. "BP and Academic Freedom." *Inside Higher Ed.* 22 July 2010. http://www.insidehighered.com/views/2010/07/22/nelson (accessed August 20, 2010)

Nelson, Michael P. and John A. Vucetich. "On Advocacy by Environmental Scientists: What, Whether, Why, and How." *Conservation Biology* 23 no. 5 (2009): 1090-1101.

Nieuwentyt, Bernard. *The religious philosopher.* Translated by John Chamberlayne. London: J. Senex et. al., 1718.

Nkansah, N., T. Nguyen, H. Iraninezhad, and L. Bero. "Randomized trials assessing calcium supplementation in healthy children: relationship between industry sponsorship and study outcomes." *Public Health Nutr.* 12, no. 10 (2009): 1931-7.

Nolen-Hoeksema, Susan. "Gender differences in depression." *Current Directions in Psychological Science* 10, no. 3 (2001): 173-176.

Normile, D., G. Vogel, and J. Couzin. "Cloning. South Korean team's remaining human stem cell claim demolished." *Science* 311, no. 5758 (2006): 156-7.

Okruhlik, Kathleen. :Gender and the Biological Sciences." *Biology and Society Canadian Journal of Philosophy Supplementary* 20 (1994): 21-42.

Oldroyd, David. *Thinking about the Earth: A History of Ideas in Geology.* London: Athlone, 1996.

Omer, Pan, Halsey, Stokley, Moulton, Navar, Pierce, Salmon, "Nonmedical exemptions to School Immunization Requirements: Secular Trends and Association of State Policies with Pertussis Incidence." *The Journal of the American Medical Association* 296 no. 14 (October 11, 2006): 1757-1763.

Osman, M. "Therapeutic implications of sex differences in asthma and atopy: Community child health, public health, and epidemiology." *Archives of Disease in Childhood* 88, no. 7 (2003): 587-590.

Ownby, D., C. Johnson, and E. Peterson. "Exposure to dogs and cats in the first year of life and risk of allergic sensitization at 6 to 7 years of age." *JAMA* 288, no. 8 (2002): 963-972.

Parker, A.J. *Ancient Shipwrecks of the Mediterranean and the Roman Provinces.* British Archaeological Reports. Oxford: Tempvs Reparatvm, 1992.

Pfeifer, Mark P. and Gwendolyn L. Snodgrass. "The continued use of retracted, invalid scientific literature." *JAMA, The Journal of the American Medical Association,* v263 n10 (March 9, 1990): 1420.

Phillips, H. "Private thoughts, public property." *New Scientist* 183, no. 2458 (2004): 38-41.

Piechota, D., R. Ballard, B. Buxton, and M. Brennan. "In situ Preservation of a Deep-Sea Wreck Site: Sinop D in the Black Sea." (forthcoming in the proceedings of the ICC 2010 Istanbul Conference: Conservation and the Eastern Mediterranean).

Pluche, Noël. *Nature Display'd.* Fouth Edition. Translated by Mr. Humphreys. London: R. Davis et. al., n.d.

Pomerleau, A., D. Bolduc, G. Malcuit, and L. Cossette. "Pink or blue: Environmental gender stereotypes in the first two years of life." *Sex Roles* 22, no. 5-6 (1990): 359-367.

Popper, Karl. "Philosophy of Science, a Personal Report." Pp. 155-191 in *British Philosophy in Mid-Century.* edited by C. A. Mace. London: Allen & Unwin, 1957.

Potter, K. M., F. W. Cubbage, G. B. Blank, and R. H. Schaberg. "A Watershed-Scale Model for Predicting Nonpoint Pollution Risk in North Carolina." *Environmental Management* (2004): 62-74.

Potter, K. M., F. W. Cubbage, and R. H. Schaberg. "Multiple-Scale Landscape Predictors of Benthic Macroinvertebrate Community Structure in North Carolina." *Landscape and Urban Planning* (2005): 77-90.

Power, Henry. *Experimental Philosophy.* Edited by Marie Boas Hall. USA: Johnson Reprint Corporation, 1966.

Preston, R. *The Hot Zone: A Terrifying True Story.* New York; Anchor, 1995.

Price, D. de S. "Gears from the Greeks: The Antikythera Mechanism—A Calendar Computer from ca. 80 B.C." *Transactions of the American Philosophical Society* (New Series) 64, no. 7 (1974): 1-70.

———. *Gears from the Greeks: The Antikythera Mechanism—A Calendar Computer from ca. 80 B.C.* New York: Science History Publications, 1975.

Quist, David and Ignacio H. Chapela. "Transgenic DNA introgressed into traditional maize landraces in Oaxaca, Mexico." *Nature* 414, no. 6863 (November 29, 2001): 541-543.

———. "Quist and Chapela reply." *Nature* 416, no. 6881 (Appril 11, 2002): 601.

Raftopoulos, A. "Was Cartesian Science Ever Meant to Be a Priori? A comment on Hatfield." *Philosophy of Science* 62, no. 1 (Mar. 1995): 150-160.

Rainey, Anson. "Historical Geography: The Link between Historical and Archeological Interpretation." *The Biblical Archaeologist* 45, no. 4 (1982): 217-223.

Rhoads, B. L., D. Wilson, M. Urban, and E. E. Herricks. "Interaction between Scientists and Nonscientists in Community-Based Watershed Management: Emergence of the Concept of Stream Naturalization." *Environmental Management* 24, no. 3 (1999): 297-308.

Richter, Daniel D. "Working Paper No. 5. Soil and Water Effects of Modern Forest Harvest Practices in North Carolina." In *Economic and ecologic impacts associated with wood chip production in North Carolina, Volume II*, Southern Center for Sustainable Forests, 2000. Http://www.env.duke.edu/scsf (accessed August 18, 2010).

Ridker, P.M. and J. Torres. "Reported outcomes in major cardiovascular clinical trials funded by for-profit and not-for-profit organizations: 2000-2005." *JAMA* 295, no. 19 (2006): 2270-4.

Robson, John. "The Fiat and Finger of God: The Bridgewater Treatises." In *Victorian Faith in Crisis: Essays on Continuity and Change in Nineteenth-Century Religious Belief*, edited by Bernard Lightman and Frank Turner, 71-125. Stanford: Stanford UP, 1990.

Rooney, Martin J. and Colleen M. Rooney. "Parental Tort Immunity: Spare the Liability, Spoil the Parent." *New England Law Review* 25 (1991): 1161-1184.

Rothman, K.J. The Ethics of Research Sponsorship. *Journal of Clinical Epidemiology* 44 (1991): 25S-28S.

———. "Conflict of interest: the new McCarthyism in science." *JAMA* 269 (1993): 2782-2784.

Royal Society. "An Abstract of Dr. Wilkins' Essay Towards a Real Character and a Philosophical Language." In *The Mathematical and Philosophical Works of the Right Reverend John Wilkins*. London: C. Whittingham, 1802.

Rusnak A.J. and A.E. Chudley. "Stem cell research: cloning, therapy and scientific fraud." *Clin Genet.* 70, no. 4 (Oct, 2006): 302-5.

Safer D.J. "Design and reporting modifications in industry-sponsored comparative psychopharmacology trials." *J Nerv Ment* Dis. 190, no. 9 (2002): 583–592.

Salmon, Daniel A., Michael Haber, Eugene J. Gangarosa, Lynelle Phillips, Natalie J. Smith, and Robert T. Chen. "Health Consequences of Religious and Philosophical Exemptions From Immunization Laws: Individual and Societal Risk of Measles." *The Journal of the American Medical Association* 282, no. 1 (July 7, 1999): 47-53.

Salmon, Merrilee. *Explanation and Archaeology*. New York: Academic Press, 1982.

———. "Explanation in Archaeology." Pp. 231-248 in *Explanation: Theoretical Approaches and Applications*, edited by G. Hon and S.S. Rakover. The Netherlands: Kluwer Academic Publishers, 2001.

Sarewitz, D. "How Science Makes Environmental Controversies Worse." *Environmental Science & Policy* (2004): 385-403.

Sarkar, Sahotra. *Biodiversity and Environmental Philosophy*. New York: Cambridge University Press, 2005.

Saul, Jennifer. *Feminism: Issues and Arguments*. Oxford: Oxford University Press, 2003.

Schaberg, Rex, Frederick W. Cubbage, and Daniel D. Richter. "Working Paper No. 2. Trends in North Carolina Timber Products Outputs, and the Prevalence of Wood Chip Mills." In *Economic and ecologic impacts associated with wood chip production in North Carolina, Volume II*, Southern Center for Sustainable Forests, 2000. Accessed August 18, 2010. Http://www.env.duke.edu/scsf (accessed August 18, 2010).

Schaberg, Rex. "Working Paper No. 11. Effects of Chip Mills on North Carolina's Aquatic Communities." In *Economic and ecologic impacts associated with wood chip production in North Carolina, Volume II,* Southern Center for Sustainable Forests, 2000. Accessed August 18, 2010. Http://www.env.duke.edu/scsf (accessed August 18, 2010).

Schaberg, Rex, P. B. Aruna, F. W. Cubbage, G. R. Hess, R. C. Abt, D. D. Richter, S. T. Warren, J. D. Gregory, A. G. Snider, S. Sherling, and W. Flournoy. "Economic and Ecological Impacts of Wood Chip Production in North Carolina: An Integrated Assessment and Subsequent Applications." *Forest Policy and Economics* (2005): 157-74.

Schiebinger, Londa. *The Mind Has No Sex? Women and the Origin of Modern Science.* Cambridge: Harvard University Press, 1989.

Schnapp, Alain. *The Discovery of the Past.* New York: Harry N. Abrams. 1997.

Schnittker, J. and G. Karandinos. "Methuselah's medicine: pharmaceutical innovation and mortality in the United States, 1960-2000." *Soc Sci Med.* 70, no. 7 (2010): 961-8.

Schroter, S., J. Morris, S. Chaudhry, R. Smith, and H. Barratt. "Does the type of competing interest statement affect readers' perceptions of the credibility of research? Randomised trial." *BMJ* 328 (2004): 742-3.

Scott, J. Michael, J. L. Rachlow, R. T. Lackey, A. B. Pidgorna, J. L. Aycrigg, G. R. Feldman, L. K. Svancara, D. A. Rupp, D. I. Stanish, and R. K. Steinhorst. "Policy Advocacy in Science: Prevalence, Perspectives, and Implications for Conservation Biologists." *Conservation Biology* 21 no. 1 (2007): 29-35.

Scriven, Michael. "Explanations Predictions and Laws." Pp. 170-230 in *Minnesota Studies in the Philosophy of Science, volume 3.* Minneapolis: University of Minnesota Press), 1962.

Sebond, Raymond. *La Theologie Naturelle de Raymond Sebon.* Translated by Michel de Montaigne. Rouen: Jean de la Mare, 1641.

Siegel M., *False Alarm: The Truth About the Epidemic of Fear.* Hoboken: Wiley, 2005.

Seltzer S.E., A. Menard, R. Cruea, and R. Arenson. "'Hyperscrutiny' of academic-industrial relationships: potential for unintended consequences—a response." *Journal of the American College if Radiol.* 7, no. 1 (2010): 39-42.

Shanks, Michael and Christopher Tilley. *Re-Constructing Archaeology.* Great Britain: Cambridge University Press, 1987.

Shapin, S. *The Scientific Revolution.* Chicago: University of Chicago Press, 1998.

Shea, W. R. *The Magic of Numbers and Motion: The Scientific Career of René Descartes.* Canton, MA: Science History Publications, 1991.

Shipley, G. *The Greek World After Alexander 323-30 BC* (The Routledge History of the Ancient World). London: Routledge 2000.

Shrader-Frechette, Kristin. "Throwing out the Bathwater of Positivism, Keeping the Baby of Objectivity: Relativism and Advocacy in Conservation Biology." *Conservation Biology* 10 no. 3 (1996): 912-914.

Sider, D. *The Library of the Villa dei Papiri at Herculaneum.* Los Angeles: J. Paul Getty Museum, 2005.

Simera, I., D. Moher, A. Hirst, J. Hoey, K.F. Schulz, and D.G. Altman. "Transparent and accurate reporting increases reliability, utility, and impact of your research: reporting guidelines and the EQUATOR Network." *BMC Med.* 8 (2010): 24.

Singer, E. "They know what you want." *New Scientist* 183, no. 2458 (2004): 36-37.

Sismondo, S. "Pharmaceutical company funding and its consequences: a qualitative systematic review." *Contemporary Clinical Trials* 29, no. 2 (2008): 109–113.

Slaughter, S., M.P. Feldman, and S.L. Thomas. "U.S. research universities' institutional conflict of interest policies." *J Empir Res Hum Res Ethics* 4, no. 3 (2009): 3-20.

Slovic, P. "Trust, Emotion, Sex, Politics, and Science: Surveying the Risk-Assessment Battlefield (Reprinted from Environment, Ethics, and Behavior, Pg 277-313, 1997)." *Risk Analysis* (1999): 689-701.

Smith, C.D. and B. MacFadyen. "Industry relationships between physicians and professional medical associations: corrupt or essential?" *Surg Endosc.* 24, no. 2 (Feb, 2010): 251-3.

Sontag S, *Illness as Metaphor and AIDS and Its Metaphors.* New York: Picador (Macmillan), 2001.

Snider, Anthony and Frederick Cubbage. "Working Paper No. 8. Nonindustrial Private Forests: an Analysis of Changes in Potential Returns as a Result of Shifts in Demand." In *Economic and ecologic impacts associated with wood chip production in North Carolina, Volume II,* Southern Center for Sustainable Forests, 2000. http://www.env.duke.edu/scsf (accessed August 18, 2010).

———. "Economic Analyses of Wood Chip Mill Expansion in North Carolina: Implications for Nonindustrial Private Forest (NIPF) Management." *Southern Journal of Applied Forestry* 30, no. 2 (2006): 102-08.

Snobelen, Stephen. "Isaac Newton, Heretic." *British Journal for the History of Science* 32 (1999): 381–419.

So, A.D., B.N. Sampat, A.K. Rai, R. Cook-Deegan, J.H. Reichman, R. Weissman, and A. Kapczynski. "Is Bayh-Dole good for developing countries? Lessons from the US experience." *PLoS Biol.* 6, no. 10 (Oct 28, 2008): e262.

Soulé, Michael. "What is conservation biology?" *BioScience* 35 (1985): 727-734.

Southern Center for Sustainable Forests. *Integrated Research Project Summary. Volume I: Economic and Ecologic Impacts Associated with Wood Chip Production in North Carolina.* 2000a, http://www.env.duke.edu/scsf (accessed August 18, 2010).

Southern Center for Sustainable Forests. *Working Papers. Volume II: Economic and Ecologic Impacts Associated with Wood Chip Production in North Carolina.* 2000b. Http://www.env.duke.edu/scsf (accessed August 18, 2010).

Spanier, Bonnie. *Im/partial Science: Gender Ideology in Molecular Biology.* Bloomington: Indiana University Press, 1995.

Sprat, Thomas. *The history of the Royal-Society of London, for the improving of natural knowledge.* London: J. Martyn, 1667.

Stamps, A.E. "Using a Dialectical Scientific Brief in Peer Review." *Science and Engineering Ethics* 3 (1997): 85-98.

Stannard, R. *Doing Away With God?: Creation and the New Cosmology.* New York: HarperCollins, 1993.

———. *The God Experiment: Can Science Prove the Existence of God?* Mahweh, NJ: Hidden Spring Books, 2000.

Steel, B., P. List, D. Lach, and B. Shindler. "The role of scientists in the environmental policy process: A case study from the American west." *Environmental Science & Policy* 7 (2004): 1–13.

Steffy, J.R. *Wooden Ship Building and the Interpretation of Shipwrecks.* College Station: Texas A&M University Press, 1994.

Sterckx, S. "Patenting and Licensing of University Research: Promoting Innovation or Undermining Academic Values?" *Science and Engineering Ethics* (Sep 19, 2009).

Stossel, T.P. "Regulating academic industry research relationships—solving problems or stifling progress?" *New England Journal of Medicine* 353, no. 10 (2005): 1060-1065.

Stump, D. "Rationalist Science" in *Blackwell's Companion to Rationalism,* edited by Alan Nelson. Blackwell Publishing, Ltd., 2005.

Slowik, E. *Cartesian Spacetime,* Dordrecht: Kluwer, 2002.

Sober, E. "Two Uses of Unification." Pp. 205-216 in *The Vienna Circle and Logical Empiricism: Re-evaluation and Future Perspectives*, edited by Stadler, F. Kluwer: The Netherlands, 2003.

Strathan, Kathleen, Alicia Gable, and Marie C. McCormick, eds. *Immunization Safety Review: Thimerosal-Containing Vaccines and Neurodevelopmental Disorder.* (Washington, D.C: National Academy Press, 2001).

Strauss, Anselm L., and Juliet M. Corbin. *Basics of Qualitative Research : Techniques and Procedures for Developing Grounded Theory.* 2nd ed. Thousand Oaks, Calif.: Sage Publications, 1998.

Suarez, Andrew V. "Conflicts around a study of Mexican crops." *Nature* 417 (2002): 897.

Summers, R.W., D.E. Elliott, J.F. Urban Jr, R. Thompson, and JV Weinstock. "Trichuris suis therapy in Crohn's disease." *Gut* 54, no. 1 (2005): 87-90.

Tavris, Carole. *The Mismeasure of Woman.* New York: Simon and Shuster, 1992.

Taylor, Walter. "A Study of Archeology." *American Anthropologist: American Anthropological Association* 50 no 3, (1948 part 2).

Temkin, Owsei and C. Lillian, eds. *Ancient Medicine: Collected Papers of Lucwig* Edelstein. Baltimore: Johns Hopkins Uniersity Press. 1969.

Thomsen, M. and D. Resnik. "The Effectiveness of the Erratum in Avoiding Error Propagation in Physics." *Science and Engineering Ethics* 1, no. 3 (1995): 231-240.

Thorne, Barrie. *Gender Play: Girls and Boys in School.* New Brunswick, NJ: Rutgers University Press, 1993.

Thucydides. *History of the Peloponnesian War.* New York: Penguin Books,1954.

Towner, J.S., T.K. Sealy, and M.L. Khristova et. al. "Newly discovered ebola virus." *PLoS Pathog* 11 no. e1000212 (Nov 4, 2008). Epub Nov 21, 2008.

Tracy, C. Richard and Peter F. Brussard. "The Importance of Science in Conservation Biology." *Conservation Biology* 10 no. 3 (1996): 918-919.

Trigger, Bruce. "Alternative Archaeologies: Nationalist, Colonialist, Imperialist," *Man* 19 (1984): 355-370.

———. *A History of Archaeological Thought.* Cambridge: Cambridge University Press, 1989.

———. "'Alternative Archaeologies' in Historical Perspective." Pp. 187-195 in *Evaluating Multiple Narratives.* New York: Springer, 2008.

Twombly, R. "Conflict-of-interest rules worry some scientists." *J Natl Cancer Inst.* 99 (2007): 6-9.

United Nations Environment Programme (UNEP) "Sixth Meeting of the Conference of the Parties to the Convention on Biological Diversity." (July 4, 2003), http://www.cbd.int/doc/?meeting=COP-06 (accessed February 18, 2009).

Valavanis, P. *Hysplex: The Starting Mechanism in Ancient Stadia: A Contribution to Ancient Greek Technology* (University of California Publications in Classical Studies) Berkeley: University of California Press: 1999.

van den Oord, EJCG. "Controlling false discoveries in genetic studies." *Am J Med Genet B Neuropsychiatr Genet.*147B (2008): 637–44.

Van Fraassen, Bas. *The Scientific Image.* Oxford: Oxford University Press, 1980.

Vassilika, E. *Greek and Roman Art.* London: Cambridge University Press: 1998.

Vechiatto, Norbert. "'Digestive Worms': Ethnomedical Approaches to Intestinal Parasitism in Southern Ethiopia." Pp. 241-266 in *The Anthropology of Infectious Disease: International Health Perspectives*, edited by Marcia Inhorn and Peter Brown. Amsterdam: Gordon and Breach Publishers, 1997.

Vogel, J., and E. Lowham. "Building Consensus for Constructive Action: A Study of Perspectives on Natural Resource Management." *Journal of Forestry* (2007): 20-26.

Voinov, A., and E. J. B. Gaddis. "Lessons for Successful Participatory Watershed Modeling: A Perspective from Modeling Practitioners." *Ecological Modelling* (2008): 197-207.

Voltaire. *Candide*. Edited by Eric Palmer, translator unknown. Peterborough & New York: Broadview Press, 2009.

Wachsmann, S. "Deep Submergence Archaeology: the Final Frontier," *Skyllis* (Deutsche Gesellschaft zur Förderung der Unterwasserarchäeologie e.V.) Proceedings of *In Poseidon's Reich* XII.8.1-2 (2007-2008): 130-154.

Wadman, M. "NIH workers see red over revised rules for conflicts of interest." *Nature* 434 (2005):3-4.

Wager, Robert, Peter Lafayette, and Wane Parrott. "Letter received on Wednesday 14, August, 2002." *Electronic Journal of Biotechnology*. http://ejbiotechnology.info/content/vol5/issue2/letters/01/index.html (accessed July 10, 2010).

Wallington, T. J., and S. A. Moore. "Ecology, Values, and Objectivity: Advancing the Debate." *Bioscience* (2005): 873-78.

Walsh S.J., and L.M. Rau. "Autoimmune diseases: a leading cause of death among young and middle-aged women in the United States." *Am J Public Health* 90, no. 9 (2000): 1463-1466.

Warren, Sarah T. "Working Paper No. 10. Social Impact Assessment: Social Impacts and community Concerns." In *Economic and ecologic impacts associated with wood chip production in North Carolina, Volume II*, Southern Center for Sustainable Forests, 2000. http://www.env.duke.edu/scsf Accessed August 18, 2010.

———. "Public Interests in Private Property: Conflicts over Wood Chip Mills in North Carolina." *Southern Rural Sociology* 19, no. 2 (2003): 114-31.

Weinfurt, K.P., M.A. Dinan, J.S. Allsbrook, J.Y. Friedman, M.A. Hall, K.A. Schulman, and J. Sugarman. "Policies of academic medical centers for disclosing financial conflicts of interest to potential research participants." *Academic Medicine*. 81, no. 2 (2006): 113–118.

Weinfurt, K.P., M.A. Hall, M.A. Dinan, V. DePuy, J.Y. Friedman, J.S. Allsbrook, and J. Sugarman. "Effects of disclosing financial interests on attitudes toward clinical research." *J Gen Intern Med*. 23, no. 6 (2008): 860-6.

White, K.D. "The Base Mechanic Arts? Some Thoughts on the Contribution of Science (Pure and Applied) to the Culture of the Hellenistic Age," Pp. 211-220 in P. Green (ed.) *Hellenistic History and Culture* (Hellenistic Culture and Society) Berkeley: University of California Press, 1996.

White, K.D. *Greek and Roman Technology*. New York: Cornell University Press, 1984.

White, L. "Technological Development in the Transition from Antiquity to the Middle Ages." *Tecnologia, economia e società nel mondo romano: Atti del Convegno di Como, 27/28/29 settembre 1979*. Como, 1980.

Whittington, C.J., T. Kendall, P. Fonagy, D. Cottrell, A. Cotgrove, and E. Boddington. "Selective serotonin receptor inhibitors in childhood depression: systematic review of published versus unpublished data." *Lancet* 363 (2004): 1341–1345.

Wilkins, John. *Of the Principles and Duties of Natural Religion*. London: Maxwell, 1675.

Wilson, J.R. "Responsible Authorship and Peer Review." *Science and Engineering Ethics* 8 (2002): 155-174.

Wilson, Leon C., Colwick M. Wilson, and Lystra Berkeley-Caines. "Age, gender and socioeconomic differences in parental socialization preferences in Guyana." *Journal of Comparative Family Studies* 34 (2003): 213-227.

Wissler, Clark. "The New Archaeology." *Natural History* 17 no. 2 (1917): 100-101.

World Association of Medical Editors (WAME). "Conflict of interest in peer-reviewed medical journals: a policy statement of the World Association of Medical Editors (WAME)." *J Child Neurol*. 24, no. 10 (2009): 1321-3.

Worobey, M, M. Santiago, and B Keele et al. "Origin of AIDS: contaminated polio vaccine theory refuted." *Nature* 428, no. 6985 (Apr 22, 2004): 820.

Worthy, Kenneth, Richard C. Strohman, and Paul R. Billings. "Conflicts around a study of Mexican crops [Correspondence]." *Nature* 417, no. 6892 (June 27, 2002): 897.

Wylie, Alison. "Reasoning About Ourselves: Feminist Methodology in the Social Sciences." Pp. 611-624 In *Readings in the Philosophy of Social Science*, edited by Martin and McIntyre. Cambridge: MIT Press, 1994. First published in *Women and Reason*, edited by E. Harvey and K. Okruhik. Ann Arbor: The University of Michigan Press, 1992.

———. "Rethinking Unity as a 'Working Hypothesis' for Philosophy: How Archaeologists Exploit the Disunities of Science." *Perspectives on Science* 7 no. 3 (1999): 293-317.

———. *Thinking From Things: Essays in the Philosophy of Archaeology*. Berkeley: University of California Press, 2002.

———. "The Integrity of Narratives: Deliberative Practice, Pluralism, and Multivocality." Pp. 201-212 in *Evaluating Multiple Narratives*. New York: Springer, 2008.

Wylie, Alison and Lynn Hankinson Nelson. "Coming to Terms with the Values of Science: Insights from Feminist Science Scholarship." Pp. 58-86 in *Value-Free Science? Ideals and Illusions*, edited by Harold Kincaid, John Dupré, and Alison Wylie. New York: Oxford University Press, 2007.

Xenophon. *Memorabilia*. Translated by H. G. Dakyns. London and New York: MacMillan & Co., 1894.

Yamamoto, Yuri T. "Values, objectivity and credibility of scientists in a contentious natural resource debate." *Public Understanding of Science*, pre published July 22, 2010, DOI: 10.1177/0963662510371435.

Young, N., and R. Matthews. "Experts' Understanding of the Public: Knowledge Control in a Risk Controversy." *Public Understanding of Science* 16, no. 2 (2007): 123-44.

Zhao G. "SARS molecular epidemiology: a Chinese fairy tale of controlling an emerging zoonotic disease in the genomics era." *Philos Trans R Soc Lond B Biol Sci.*; 362, no.1482 (June 29 2007): 1063–1081.

Zvidi, I., R. Hazazi, S. Birkenfeld, and Yaron Niv. "The prevalence of Crohn's disease in Israel: A 20-year survey." *Digestive Diseases and Sciences* 54, no. 4 (2008): 848-852.

Zuany-Amorim, C. et. al. "Suppression of airway eosinophilia by killed Mycobacterium vaccae-induced allergen-specific regulatory T-cells." *Nat. Med.* 8 (2002): 625–629.

Zycher, B., J.A. DiMasi, and C.P. Milne. "Private sector contributions to pharmaceutical science: thirty-five summary case histories." *American Journal of Therapy* 17, no. 1 (2010): 101-20.

# Index

# About the Contributors

**Evelyn Brister** is assistant professor of philosophy at Rochester Institute of Technology. Her research in epistemology and the philosophy of science examines the role of social structures in supporting knowledge claims. In particular, her current work analyzes how unrecognized epistemological, scientific, and social assumptions are used to justify conservation and land management policies. As part of this work she has collaborated with plant biologists and environmental scientists to conduct an historical survey of western New York forests.

**Bridget Buxton** is assistant professor of history at the University of Rhode Island. She received her Ph.D. in ancient history and Mediterranean archaeology from the University of California, Berkeley, and holds an MA (distinction) and BA hons. in Classical Studies from Victoria University, New Zealand. She has been involved in numerous archaeological projects in the Mediterranean on land and underwater (with the Institute for Nautical Archaeology), and very deep underwater in her collaborations with Prof. Robert Ballard of the Institute for Exploration and Prof. Shelley Wachsmann of Texas A&M Nautical Archaeology Program. She is a NAUI assistant dive instructor and AAUS scientific diver, and lectures about underwater archaeology for the Archaeological Institute of America and on cruise ship circuits.

**Sharyn Clough** is associate professor of philosophy at Oregon State University. She uses a contemporary pragmatist approach to semantics as a tool to investigate the role of feminist and other values in scientific practice. She is the author of *Beyond Epistemology: A Pragmatist Approach to Feminist Science Studies* (Rowman and Littlefield 2003), and the editor of *Siblings Under the Skin: Feminism, Social Justice and Analytic Philosophy* (Davies Group 2003). In addition, she has written a number of essays and reviews for journals such as *Studies in the History and Philosophy of the Biological and Biomedical Sciences, Social Epistemology, Social Philosophy* and *Hypatia*. She can be contacted at: sharyn.clough@oregonstate.edu.

**Steven C. Hatch** is instructor in the Department of Medicine and the University of Massachusetts Medical School, Worcester, MA. He works in the Division of Infectious Diseases and Immunology as a researcher on the pathogenesis of Dengue Hemorrhagic Fever. He received his MD at the University of Cincinnati School of Medicine, did his residency in Internal Medicine at Tufts University—

New England Medical Center, and specialized in Infectious Diseases at UMass, where he has stayed on as faculty. He is the author of *Blind Man's Marathon*, a memoir about his medical school experiences.

**Kristen Intemann** is assistant professor of philosophy at Montana State University. Her research focuses on the roles of ethical, social, and political values in scientific research and the implications this has for scientific practices and conceptions of objectivity. She has published in journals including *Philosophy of Science, Biology & Philosophy, Hypatia, Social Epistemology, FASEB Journal,* and *The European Journal of Epidemiology*.

**Brian L. Keeley** is professor of philosophy at Pitzer College in Claremont, California, where he teaches in the Philosophy, Neuroscience and Science, Technology & Society Programs. In addition to work on religion and science, he has also published work in the philosophy and cognitive science of the sensory modalities, the philosophy of biology, and the strange epistemology of conspiracy theories. He also edited *Paul Churchland*, a collection of papers on the eponymous philosopher of mind and science.

**William H. Krieger** is assistant professor of philosophy at the University of Rhode Island. He serves as the Executive Secretary of the International Society for the History of Philosophy of Science and has served on regional boards of ASOR and SBL. He has been involved in a variety of archaeological projects since 1994 and is currently co-director of the Israel Coast Exploration Project, a combined land-sea archaeological excavation focusing on Israel's Hellenistic coastline. His latest book, *Can There be a Philosophy of Archaeology* (Lexington Press), focuses on the connection between the history and philosophy of science and archaeological theory and methodology. Correspondence should be addressed to: krieger@mail.uri.edu.

**Inmaculada de Melo-Martín** is associate professor in the Division of Medical Ethics, Weill Cornell Medical College. She holds a Ph.D. in Philosophy and a M.S. in Molecular Biology. Her research focuses on ethical and epistemological issues related to biomedical science and technology. She is the author of *Making Babies* (Kluwer, 1998) and *Taking Biology Seriously* (Rowman and Littlefield, 2005).

**Eric Palmer** is associate professor of philosophy at Allegheny College, Pennsylvania. His writing in science studies considers philosophical aspects of science from the seventeenth to the mid-eighteenth century, with foci upon French philosophy of science from Descartes to Voltaire. Eric has edited the Broadview Critical Edition of *Candide*. He is currently examining the intellectual context of physico-theology across Europe, and the paper in this volume is the first fruit of that effort. Eric also thinks and writes at the intersection of business ethics, development ethics and economic globalization. Various work may be accessed at: http://webpub.allegheny.edu/employee/e/epalmer/paperfile/papers.html

**Anya Plutynski** is associate professor of philosophy at the University of Utah. Her research has been primarily in history and philosophy of biology. Her research interests are in early twentieth century evolutionary theory, scientific explanation, modeling, unification, and reduction, with publications in the *Proceedings of the Philosophy of Science Association*, the *British Journal for the Philosophy of Science*, *Studies in the History and Philosophy of the Biological Sciences*, and *Biology and Philosophy*, and co-editing *Blackwell's Companion to Philosophy of Biology*.

**Adam D. Roth,** Ph.D., is assistant professor of communication studies at The University of Rhode Island. Correspondence about this paper can be addressed to the author at: adamroth@uri.edu.

**Glenn Sanford** is associate professor of philosophy at Sam Houston State University. He holds a Ph.D. in philosophy of biology from Duke University and is completing a J.D. at the University of Houston. He has published articles with biologists, historians, and psychologists on topics ranging from the use of the comparative method in evolutionary modeling and the history of the creation/evolution debates to critical analysis of purported evolutionary explanations of psychopathy. His current research interest, the impetus for pursuing his law degree, is the interaction of science and democracy within law and public policy.

**Lawrence Souder** earned his Ph.D. in the rhetoric of science from Temple University. He is on the graduate faculty of culture and communication at Drexel University. Souder's research is focused on the ethics of the scholarly peer review process. His most recent paper, "A rhetorical analysis of apologies for scientific misconduct: Do they really mean it?" appeared in *Science and Engineering Ethics*.

**Yuri Yamamoto** is visiting researcher at the Department of Forestry and Environmental Resources at North Carolina State University. From 2003 to 2004 she worked in the Department of Agricultural and Resource Economics at North Carolina State University with a National Science Foundation Professional Development Fellowship [NSF grant No. 0328433] to learn social science methods and to study experiences of participants of North Carolina Wood Chip Production Study. She is trained as a plant molecular geneticist and is interested in how scientists can work with citizens to inform natural resource policy. Her research standpoint is one of a sympathetic academic colleague to the scientists rather than that of a detached observer. Yuri was a participant of the North Carolina Natural Resource Leadership Institute where she first learned about the North Carolina Wood Chip Production Study. She also participated in a U.S. Geological Survey workshop on joint fact-finding and the National Endowment for the Humanities Science and Values Institute at University of Pittsburgh in 2003, which gave her invaluable experience to interact with philosophers of science. Correspondence should be addressed to: yuri@ncsu.edu.